Ethical, Social and Psychological Impacts of Genomic Risk Communication

This volume presents the ethical implications of risk information as related to genetics and other health data for policy decisions at clinical, research and societal levels.

Ethical, Social and Psychological Impacts of Genomic Risk Communication examines the introduction of new types of health risk information based on faster, cheaper and larger sets of genetic or genomic analysis. Synthesizing the results of a five-year interdisciplinary project, it explores the unsolved ethical and social questions around the sharing of this data, such as: What is best practice in risk communication? What are the normative presumptions and ethical consequences of an increased individual responsibility for ones' health? And how does one deal with the gap between the knowledge of risk and the lack of therapeutic options which often exist for complex diseases, such as dementia or some types of cancer? Drawing on contributions from over 20 experts in the field, this collection examines these questions from a liberal bioethics' perspective, advocating for contextual and cultural-sensitive ethical discussions.

This book will be of great interest to students and scholars of theoretical and clinical medical ethics, medical sociology, risk communication and ethics of risk, as well as professionals in clinical genetics.

Ulrik Kihlbom is Associate Professor in Medical Ethics at Uppsala University.

Mats G. Hansson is Senior Professor of Biomedical Ethics and Director of The Centre for Research Ethics & Bioethics (CRB) at Uppsala University.

Silke Schicktanz is Full Professor of Cultural and Ethical Studies of Biomedicine at University Medical Center Göttingen.

Earthscan Risk in Society series
Edited by Ragnar E. Löfstedt
King's College London, UK

Anthropology and Risk
Åsa Boholm

Explaining Risk Analysis
Protecting health and the environment
Michael R. Greenberg

Risk Conundrums
Solving Unsolvable Problems
Edited by Roger E. Kasperson

Siting Noxious Facilities
Integrating Location Economics and Risk Analysis to Protect Environmental Health and Investments
Michael R. Greenberg

Moral Responsibility and Risk in Society
Examples from Emerging Technologies, Public Health and Environment
Jessica Nihlén Fahlquist

Risk and Uncertainty in a Post-Truth Society
Edited by Sander van der Linden and Ragnar E. Lofstëdt

A Pre-Modern Cultural History of Risk
Imagining the Future
Gaspar Mairal

Ethical, Social and Psychological Impacts of Genomic Risk Communication
Edited by Ulrik Kihlbom, Mats G. Hansson and Silke Schicktanz

For more information about this series, please visit: www.routledge.com/Earthscan-Risk-in-Society/book-series/ERSS

Ethical, Social and Psychological Impacts of Genomic Risk Communication

Edited by Ulrik Kihlbom,
Mats G. Hansson and
Silke Schicktanz

Routledge
Taylor & Francis Group
LONDON AND NEW YORK

earthscan
from Routledge

First published 2021
by Routledge
2 Park Square, Milton Park, Abingdon, Oxon OX14 4RN

and by Routledge
52 Vanderbilt Avenue, New York, NY 10017

Routledge is an imprint of the Taylor & Francis Group, an informa business

British Library Cataloguing-in-Publication Data
A catalogue record for this book is available from the British Library

Library of Congress Cataloging-in-Publication Data
Names: Kihlbom, Ulrik, editor. | Hansson, Mats G., editor. | Schicktanz, Silke, editor.
Title: Ethical, social and psychological impacts of genomic risk communication / edited by Ulrik Kihlbom, Mats G. Hansson, and Silke Schicktanz.
Description: Abingdon, Oxon ; New York, NY : Routledge, 2021. | Includes bibliographical references and index.
Identifiers: LCCN 2020026076 (print) | LCCN 2020026077 (ebook) | ISBN 9780367356699 (hardback) | ISBN 9780429341038 (ebook)
Subjects: MESH: Health Communication–ethics | Genomics–ethics | Risk Factors | Health Information Management–ethics | Consumer Health Information–trends
Classification: LCC R724 (print) | LCC R724 (ebook) | NLM WA 590 | DDC 174.2–dc23
LC record available at https://lccn.loc.gov/2020026076
LC ebook record available at https://lccn.loc.gov/2020026077

ISBN: 978-0-367-35669-9 (hbk)
ISBN: 978-0-429-34103-8 (ebk)

Typeset in Sabon
by Wearset Ltd, Boldon, Tyne and Wear

Contents

Notes on contributors vii
Acknowledgements xii

1 **Introduction** 1
ULRIK KIHLBOM, MATS G. HANSSON AND SILKE SCHICKTANZ

PART I
New visions of 'genetics and genomics' as integrative medicine 11

2 **Genomic medicine in 2025–2030** 13
MARTINA C. CORNEL AND GERTJAN VAN OMMEN

PART II
Risk empowerment 25

3 **Genomic risk perception and implications for patient outcomes from genetic counselling** 27
MARION MCALLISTER

4 **Genomic and biological risk profiling: from medicalization to empowerment** 35
MATS G. HANSSON, KARIM RAZA AND MARIE FALAHEE

PART III
Rethinking professional responsibility 49

5 On patients' difficulties in understanding medical risks and the aims of clinical risk communication: 'They don't really understand' 51
ULRIK KIHLBOM

6 Ethical consideration about health risk communication and professional responsibility 62
SILKE SCHICKTANZ

7 Genetic testing and risk perception in the context of personalized medicine: how familial relationships matter? 79
SABINE WÖHLKE, MARIE FALAHEE AND KATHARINA BEIER

PART IV
Social and policy implications 99

8 Direct to consumer personal genomic testing and trust: a comparative focus group study of lay perspectives in Germany, Israel, the Netherlands and the UK 101
MANUEL SCHAPER, AVIAD RAZ, MARIE FALAHEE, KARIM RAZA, DANIELLE TIMMERMANS, ELISA GARCIA GONZALEZ, SILKE SCHICKTANZ AND SABINE WÖHLKE

9 The application of new technologies to improve literacy among the general public and to promote informed decisions in genomics 130
SERENA OLIVERI, RENATO MAINETTI, ILARIA CUTICA, ALESSANDRA GORINI AND GABRIELLA PRAVETTONI

10 Algorithms as emerging policy tools in genomic medicine: opportunities and challenges ahead 148
FREDERIC BOUDER

11 Epilogue 159
MATS G. HANSSON, ULRIK KIHLBOM AND SILKE SCHICKTANZ

Index 175

Contributors

Katharina Beier is a Research Associate at the Department of Medical Ethics and History of Medicine at the University Medical Center Göttingen and the Head of the Office of Ombuds Matters and Good Scientific Practice at Göttingen University. In her research, she focuses on research ethics, particularly with regards to biobanking and big data-based research, as well as on ethical issues of assisted reproduction. In this latter context, she has published on understandings of family and parenthood, reproductive autonomy and practices of third-party reproduction, such as surrogacy.

Frederic Bouder is Full Professor in Risk Management at University of Stavanger. For over 15 years he has developed research on risk policy and risk communication, with a strong European and transatlantic dimension. Frequently, his research has focused on critical decisions that fall within a grey area where the benefits for society are clearly recognizable but where simultaneously the risks are perceived as significant. His research has been primarily developed in the context of Pharmaceuticals as well as Food, Environment and Health & Safety. Bouder has built strong collaborations with regulatory agencies across Europe and North America. He has edited four volumes and published over 40 peer-reviewed articles and book chapters.

Martina C. Cornel is Professor of Community Genetics and Public Health Genomics at Amsterdam University Medical Centres, Amsterdam, The Netherlands. She is principal investigator in the research programs Amsterdam Public Health and Amsterdam Reproduction and Development. Her research interests are responsible innovation in relation to new genetic/genomic techniques and population screening. She is a member of the Health Council of the Netherlands. She is the author of more than 200 peer-reviewed papers.

Ilaria Cutica is Associate Professor in General Psychology at the Department of Oncology and Hemato-Oncology of the University of Milan, Italy. She teaches courses on Cognitive Psychology. Her research interests focus principally on patients' empowerment, patients' decision processes, and communication in medical settings. She is author of more than 30 peer review papers.

Marie Falahee is a Lecturer in Behavioural Rheumatology at the Institute of Inflammation and Ageing, University of Birmingham, United Kingdom. Her research focuses on patient preferences for predictive/preventive strategies and therapeutic interventions for rheumatoid arthritis. She also publishes research on strategies to reduce treatment delay for rheumatoid arthritis, patient involvement in rheumatology research and the use and assessment of patient reported outcomes for inflammatory diseases.

Elisa Garcia Gonzalez has a PhD in Molecular Biology as well as one in Medical ethics. She currently works at University Medical Center – Vrije Universiteit Amsterdam. Her research interests are genomic prenatal screening and the ethics of human germline modifications. Previously, she worked in the European project Precedi (Personalized PREvention of Chronic DIseases) aiming to promote a proper integration of –omics information into public health interventions and will collaborate with its follow project EXACT.

Alessandra Gorini is a clinical psychologist and a researcher at the Department of Oncology and Hemato-Oncology, University of Milan. She is also the head of the Psycho-Cardiology Unit of Centro Cardiologico Monzino, Milan. Her research interests are mainly focused on the link between cognitive and emotional mental aspects and cardiovascular and oncologic diseases, and the role of technology in health care. She has authored more than 100 publications, including over 60 publications in international indexed journals, numerous book chapters and two monographs.

Mats G. Hansson is Professor of Biomedical Ethics and Director of The Centre for Research Ethics & Bioethics (CRB) at Uppsala University. The Centre is a multi-disciplinary unit awarded for international excellence with long experience of managing large international projects, currently leading research on risk conceptualization and communication. This 30-person team of researchers is combined with social scientists, psychologists and risk communication experts, as well as ethicists for assessment of individual preferences and public health needs and frontline research in different disease areas. Philosophers, medical doctors, nurses and lawyers are also part of the team.

Ulrik Kihlbom is Associate Professor in Medical Ethics at Uppsala University. He has a PhD in practical philosophy from Stockholm University and has normative ethics, methodology of bioethics, clinical ethics and decision-making as research interests. He is currently academic co-lead for WP2 in the IMI-PREFER project and a member of the scientific advisory committee of the Swedish Agency for Health Technology Assessment and Assessment of Social Services. He has published two books and more than 30 peer-reviewed papers.

Renato Mainetti is a research fellow at the Department of Computer Science, University of Milan and collaborates with the Applied Intelligent Systems Laboratory. He does research in the fields of real time video/image analysis, artificial intelligence and serious games; he teaches AI and Virtual Reality. His research focuses on the development of Games for Rehabilitation (Neglect, Post stroke, Haemophilia, Multiple Sclerosis and Congenital hand anomalies) and Games for Learning about genetics. He has published more than 30 peer-reviewed papers in the field of serious games.

Marion McAllister is a registered genetic counsellor (GCRB and EBMG), with a PhD in Social Science from Cambridge University (1999), and currently Reader and Programme Director for the MSc in Genetic and Genomic Counselling at Centre for Medical Education, Cardiff University, UK. She has a special interest in patient empowerment, and a distinguished track record in genetic counselling research. Her research aims to improve how clinical genetics services respond to the needs of families with genetic conditions, including addressing issues relating to patients' understanding and perceptions of genetic risks. She developed and validated a new Patient Reported Outcome Measure (Genetic Counselling Outcome Scale GCOS-24) for clinical genetics services, which captures patient benefits conceptualized as patient empowerment. She has also published research on patient empowerment in chronic disease. Since publication in 2011, GCOS-24 has been translated into Dutch, Danish, Portuguese, Spanish and Japanese and is being used to evaluate clinical genetics services in practice, and new interventions in research in the UK and internationally.

Serena Oliveri is a Post-Doc researcher in Cognitive Psychology and Decision-Making processes at the at the Department of Oncology and Hemato-Oncology, University of Milan and at the Applied Research Division for Cognitive and Psychological Science at the European Institute of Oncology (IEO). She does research in medical decision-making and patient empowerment. Her research interests mainly focus on the psychological implication of genetic testing and genetic risk information, educational tools in genetics, mHealth technologies and supportive care in cancer patients. She has published one book and has authored two chapters in edited volumes and more than 30 peer-reviewed papers.

GertJan van Ommen is Prof. em. Dr. van Ommen and has headed the Department of Human Genetics of Leiden University Medical Center (LUMC) from 1992–2012. His research interests are the application of genomics technology in basic and translational study of rare and common disease. He established the Leiden Genome Technology Center (LGTC, a national genomics facility), and the Center for Medical Systems Biology (CMSB, a genomics research/infrastructure consortium between six institutions in Amsterdam, Leiden and Rotterdam). In parallel, he has a great interest in ELSI aspects of genome science and in public and professional communication. He is Editor-in-chief of the European Journal of Human Genetics, past president of HUGO (1998–2000) and of the European (2003–2004) and Dutch (1993–2000) Societies of Human Genetics. He was National Coordinator for Ophanet, founding member of the European Biobanking and Biomolecular Research Infrastructure BBMRI-ERIC and of BBMRI-NL (www.bbmri.nl), which he directed 2009–14. His group pioneered the exon-skipping approach for therapy of Duchenne Muscular Dystrophy, currently FDA-approved. He has published circa 600 papers and supervised 50 PhD candidates.

Gabriella Pravettoni is Full Professor of Psychology of Decision Making at the University of Milan, Director of the Psychoncology Division at the European Institute of Oncology and Director of the Department of Oncology and Haemato-oncology at the University of Milan. Her research activities are in the field of cognitive sciences, psychoncology and personalized medicine. Her main interest concerns medical and clinical decision-making, patient empowerment and well-being within the paradigm of the humanization of care. She has published 126 peer reviewed papers, more than 40 books and 150 abstracts.

Aviad Raz is full Professor of Medical and Organizational Sociology at the Department of Sociology and Anthropology, Ben-Gurion University, Israel, where he is Chair of the M.A. Program. He was a visiting Professor at the Department of Sociology, University of California, San Diego, and a visiting scholar at SciencesPo and the University of Leuven. Raz has written eight books and over 70 peer-reviewed articles and chapters on topics in medical and organizational sociology and bioethics. His recent books include *Comparative Empirical Bioethics: Dilemmas of Genetic Testing and Euthanasia in Israel and Germany* (2016, with Silke Schicktanz) and *Cousin marriages: between tradition, genetic risk and cultural change* (2015, with Alison Shaw).

Karim Raza is Versus Arthritis Professor of Rheumatology at the University of Birmingham and Honorary Consultant Rheumatologist at Sandwell and West Birmingham Hospitals NHS Trust, Birmingham, UK. His research focuses on rheumatoid arthritis, in particular its earliest stages, and addresses disease mechanisms, predictors of outcome and patient perspectives on their disease and its treatment. He has published more than 160 peer reviewed papers and written ten book chapters.

Manuel Schaper is an associate at the Department of Medical Ethics and History of Medicine at the University Medical Center Göttingen. His main interests and publication focus are ethical aspects of predictive genetic diagnostics, ethical aspects of public health communication and implications of commercialization in health care. In his work he applies qualitative social-empirical methods in an empirical-ethical research approach.

Silke Schicktanz is Full Professor of Cultural and Ethical Studies of Biomedicine at the University Medical Center of Göttingen, Germany. She does research and teaches in the interdisciplinary field of bioethics and cultural studies of medicine. Her research focuses on cross-cultural bioethics, lay-expert-interaction, and concepts of responsibility in various fields of modern medicine (dementia, transplantation, genetics, etc.). She has published two books, ten edited volumes and more than 60 peer review papers.

Danielle Timmermans is Full Professor at the Department of Public and Occupational Health, University Medical Center – Vrije Universiteit Amsterdam. Her primary research interests are risk perception, risk communication and decision-making. She has published extensively on these topics. She is head of the interdisciplinary research group RISC Amsterdam. This research group is internationally prominent and in the Netherlands is the leading research group on perception and communication of health risks. Research is on perception and communication with the public and patients about health and safety risk in order to foster informed judgement and deliberate decision-making about, e.g., prenatal screening, cancer screening, cancer treatment, vaccination, environmental risks in collaboration with societal and professional partners (e.g. GGD, RIVM).

Sabine Wöhlke is Research Associate at the Department of Medical Ethics and History of Medicine at the University Medical Center Göttingen. Her main interests are ethical and cultural aspects of personalized medicine, e.g. predictive genetic testing and biomarker research. She used for her research an empirical-ethics approach and has experience with socio-empirical research in the field of patient–physician communication and shared decision-making. She also has cultural comparative experience and is currently working on several comparative studies. In this latter context, she has published on responsibility and genetic risk, decision-making processes in medicine, physician–patient communication and on cultural differences in several medical ethical fields.

Acknowledgements

This work was supported by the **Riksbankens Jubileumsfond** (The Swedish Foundation for Humanities and Social Sciences), Project *Mind the Risk* under Grant No. 1351730.

Frederic Bouder wishes to thank Sanja Mrksic Kovacevic for the continuous support and advice that she provided.

Manuel Schaper et al. wish to thank Mark Schweda, Alexander Urban, Julia Perry, Yoav Reisner, Imogen Wells, Gwenda Simons and all the focus group participants. Manuel Schaper has received additional support by the Graduate School of Humanities Göttingen (GSGG completion grant for doctoral candidates). Karim Raza is supported by the NIHR Birmingham Biomedical Research Centre, University Hospitals Birmingham NHS Foundation Trust and University of Birmingham.

1 Introduction

Ulrik Kihlbom, Mats G. Hansson and Silke Schicktanz

Background: a shift from old to new risk information in health care

Traditionally, genetic testing was confined to specialist medical genetic services, focused on relatively rare, high penetrance inherited diseases. In the last three decades, this type of genetic testing has triggered an intense and rapidly growing international debate on the ethical, legal and social aspects. However, the most common and complex disorders such as heart diseases, diabetes, arthritis, cancer and age-related dementia are usually the result of variation in many genes, each contributing a small amount of genetic susceptibility, acting in concert with lifestyle, environmental or epigenetic factors. Some of these external, non-genetic factors might be modifiable (such as stopping smoking, better nutrition, more exercise, reducing alcohol intake). Therefore, knowing that one is genetically at a higher risk might give individuals reason to avoid those manageable factors to counterbalance their risk. This marks an important difference between 'old (mono)genetic risk' and new multi-factorial health risk. Various technologies, including whole genome sequencing, testing of genetic susceptibilities, big data, epidemiology, in combination with other lifestyle-related information, mean that a new type of risk information is now available, even if still often only applied in the research context. It will impact – or should we say transform – our health care in the near future. The future vision of this new medicine is genomic, personalized, risk-predictive and preventive (see Chapter 2).

However, the introduction of this new type risk information raises important, unsolved ethical and social questions: How to best do risk communication? What are the normative presumptions and ethical consequences of an increased individual responsibility for ones' health? And how to deal with the gap between the knowledge of risk and the lack of therapeutic options for some complex diseases, such as dementia or some types of cancer? Another challenge is how to manage a development where new diagnostic tests regarding genetic risk are introduced in health care without a sufficient degree of accuracy. Different prediction models for estimation of a risk are typically evaluated on the basis of their capacity to discriminate between events and non-events, e.g. recurrence of a cancer after surgery or diagnosis of a cancer. These models may be accurate in providing good estimates of false-positive and false-negative diagnostic results, but these measures will not provide information about the clinical consequences. A false-negative result may be more harmful than a false-positive since a false-negative will falsely give the impression that one is not going to get cancer. Decision analysis will need measurement and valuation of different health outcomes, such as quality of life, life-years saved, etc. In cases where tests are used to guide treatment decisions more directly and when, e.g., the likely benefits of some treatment are not much larger or much smaller than another treatment, such effects of a treatment may be large enough for some patients to decide for treatment A, but not for others. It will depend also on their attitudes towards risk. Clinical decision-making should always be made in close collaboration with the patient. However, in such particular cases shared decision-making requires patients to be well informed by their attitudes and preferences related to risk as well as the psychosocial effects of risk information.

This volume is one of the synthesizing results of the interdisciplinary five-year long European research program *Mind the Risk*, funded by the Swedish Foundation for Humanities and Social Sciences. The overall purpose of the research program was to identify ethical issues, clarify philosophical concepts and to find out and clarify the consequences and ethical challenges of new genetic techniques through various empirical approaches. It was clear from the outset that a broad multidisciplinary approach was needed to tackle these issues. Collaboration began with psychologists and game developers in Milan, philosophers and ethicists in Göttingen and Uppsala, health economists in Manchester, rheumatologists and clinical psychologists in Birmingham, risk sociologists in Nijmegen, as well as clinical geneticists and midwives in Stockholm. Nearly 30 researchers, both senior and junior, have been involved in the project. This volume focuses on some of the main results of this project through contributions that elaborate the project's ethical and social dimensions.

Risk understanding as key for communication and decision-making

New genetic risk has a double structure: on the one hand, there is individual health risk connected to genetic information, as well as individual health behaviour and on the other hand, there are social aspects of public health. The individual meaning ascribed to health risks as well as the normative evaluation of them on a societal level can vary broadly (see also Chapter 3). Individual interpretation does not necessarily match scientific risk assessment. The evaluation of risk varies between persons but also with regard to time and place: anticipating a possible future event can differ significantly from the evaluation of the same event that has already taken place (see also Chapter 5). Furthermore, cultural and social factors have an impact on the evaluation of adverse events and risk and need to be taken into account when discussing risk communication. The identification and analysis of such socio-cultural aspects require, however, difficult comparative research settings (see Chapter 8). Besides the analysis of factors influencing the individual perception of risk and possible coping strategies, a philosophical analysis of risk and risk communication also needs to take the different theoretical perspectives of risk into account. Risk and uncertainty are distinctly different. Risk is linked to the knowledge of the probability of possible outcomes, whereas under uncertainty the probability is unknown. However, in medical contexts there is rarely a thorough knowledge of probabilities assigned to the different outcomes. The term risk is often used even when probabilities are unknown and the theoretical distinction between risk and uncertainty becomes in everyday settings blurred. Risk communication not only has to deal with the already complex statistical information of probabilities and different ways to calculate risk. Most importantly, the problem of uncertainty and of trust (or mistrust) – as a reaction to the uncalculated uncertainty – in scientific knowledge about these probabilities, as well as the trust in the clinical practice of handling risks, need to be considered. Moreover, there are also competing forms of knowledge production and value assessments. This is especially true in the field of genetic risk, where not only single agents are acting, but social groups such as professions, patient collectives, political stakeholders or private companies. Whereas some risks are part of individual risk behaviour – like smoking – some risks can potentially affect anybody, as the new pandemic COVID-19 taught us in 2020. However, some persons might be more biologically vulnerable due to their genetic, physical or social conditions.

Policy making, in turn, must manage conflicting views and ideologies on risk, health and social values, such as freedom or privacy. On the one hand, the increasing amount of genetic and health information available can be promoted as allowing more individual liberties, as it is the individual who has to evaluate and choose. On the other hand, the intense data collection related to this trend is opening the door for intrusions into individual and family privacy. Health care systems seem yet largely unprepared for how to communicate complex risk information in a way that takes values, beliefs, psychological needs and balanced preferences of individuals and families into consideration. Furthermore, the transnational movement of patients and worldwide offers of direct-to-consumer health services via internet platforms make it necessary to develop approaches that go beyond the traditional understanding of national or local regulation. Here, we are just in an early phase of increasing awareness and there is a need for more discussions and framework development. Therefore, international and interdisciplinary exchanges on the ethical and social implications become increasingly important in order to develop a cross-cultural understanding of these life sciences challenges for the individual and societies.

This is where this volume steps in; the compiled work will bring out some major ethical implications of risk information related to genetics and other health data for policy decisions at clinical, research and societal levels. By systematically combining work from theoretical and empirical ethics, we will further elaborate upon the theoretical aspects underlying the assumptions of health risk informed by various sources of modern life science research, integrating genetics, protein-markers and epidemiological information.

The need for empirical ethics to examine the ethical issues of genetic and health risk

This volume is exceptional by showing systematically how socio-empirical research with lay persons, experts and patients can substantially contribute to the international ethical debate. By this, we aim at more contextual and cultural-sensitive ethical discussions. We briefly methodologically reflect on this empirical approach, as it is not an uncontroversial move within bioethics in terms of three closely interrelated claims.

The first claim is that of uncodifiability of morals. This is the idea that correct/valid ethical norms and values cannot be fully captured in finite universal ethical principles. Considerations that ethically favour a certain course of action here in context A, may not function in the same way there, in context B. The claim of uncodifiability goes back to Aristotle, and can be found in the tradition of Casuistry, the works of the later Wittgenstein, as well as in more recent literature on moral particularism. Particularism also parallels recent developments in debates over the role of judgement in professional ethics and medical epistemology. As a fundamental claim about the structure of ethical norms, the uncodifiability claim lays the ground for the next two methodological claims.

The second claim is that bioethics should be naturalized and is here to be understood as the claim that the relevant ethical problems and the tangible ethical values and norms must be identified where they occur – in the actual practices. This means that ethicists or philosophers cannot, *a priori*, state what the relevant ethical problems, norms or values are in health care or in any other field of practice. The particular ways in which ethical quandaries, values and norms may surface in a practice such as health care can be known *a posteriori*. Careful empirical investigations with multi-disciplinary teams should provide this input to critical and ethical discussions. For instance, one may need the help of psychologists examining the effects of genetic risk information and health preference economists examining: i) what matters to patients; ii) how much it matters and; iii) what matters most to patients, when it comes to patients' trade-offs between benefits and risks. For this reason, it is also important that the ethical perspective is present when designing the empirical study.

Lastly, there is a role for critical perspectives and ethical judgement in the approach indicated here. It can, but need not be, an overall ethical judgement phrased in terms of morally right or wrong. It may also be a judgement that some ways to state the ethical problems are misleading or tend to miss what people experience as important. Against the backdrop of the claims above, it is perhaps clear that such a perspective or normative judgement would be contextual. Even if limited in scope, it would be evidence-based in a way that many bioethical inquiries are not.

Overview of the volume's structure and content

Traditional clinical ethics addressing ethical issues of communication tends to focus on the clinical setting of doctor–patient relationship. All existing guidelines (UNESCO, WMA, CIOMS, Council of Europe) mainly address professionals within the health care sector. This remains still valid in the new paradigm of more personalized, more precise and more stratified medicine. However, the social levels, including families, patient organizations or health policy organizations are often overlooked. Studies from public health ethics as well as sociology have pointed out this gap in liberal bioethics. The new perspectives to the existing corpus of 'genetics and ethics' literature we want to introduce, can be categorized along four main themes: new visions of 'genetics and genomics' as integrative medicine; risk empowerment; rethinking professional responsibility; and social and policy implications.

1. New visions of 'genetics and genomics' as integrative medicine: the first is looking ahead and broadening the perspective of genetics. Academic work in bioethics and genetics has in the past focused very much on mono-genetic testing at the beginning of life (e.g. such as prenatal or preimplantation diagnostics). Another area of bioethical literature has mainly problematized the scientific validity of genetic testing for complex, multifactorial diseases. However, this last field of genetic testing is exactly what is part of the new paradigm of personalized – or better – stratified – medicine in the twenty-first century. It relies on new forms of high-throughput genomic analysis and focuses on many genes or whole genomes. Furthermore, and no less importantly, it embeds genomic information in other health risk information (e.g. lifestyle, age, social background, ethnicity, non-genetic parameters such as body weight, blood tests, and physical examinations) and by this tries to overcome the former simplistic ideas to use monogenetic risk information to target multifactorial as well as rare diseases in a way that blurs the distinction between prevention and treatment.

Therefore, we will start this volume with the visionary, but not unrealistic, outlook of two experts from the field: Martina C. Cornel and GertJan van Ommen will discuss how genomic medicine is finding its place in mainstream medicine. Through the example of oncology, they show how whole genome sequencing helps to determine the best tumour targets and optimal treatment. Also, how prognostic risk assessment and integrating polygenic risk scores (PRS) will become part of public health programs while the testing is getting cheaper and more affordable. More nations will establish data banks, and whole genome sequencing repositories will link these with disease records, biobank data and pharmacogenetics insights. They will also discuss how limited resources and quality standards determine to what extent this will come to fruition and how to overcome health inequity. Hence, patients and citizens have not only access to their own medical records, but increasingly also access to real-time health data, monitored by home devices and wearables. Responsibilities are then shared between physicians and citizens. While more therapies for hundreds of rare diseases are available, modalities of reimbursement might be unsettled. So according to their vision, the boundary between screening and diagnostics will fade away, and health care and public health will converge into genomic medicine.

2. 'Risk empowerment': this new medical vision will be accompanied by new approaches to the concept of empowerment and its relevance for genomic risk communication. Following rapid progress in genome sequencing, genetic information will to an increasing degree be relevant in clinical settings in order to provide more precise and personalized diagnosis and treatment for patients. However, genetic diseases distinguish from chronic diseases in the sense that many are not treatable, they influence entire families to a higher extent, patients include healthy at-risk relatives and they are connected to life-changing decisions. Risk information includes unsecure future outcomes which can make internal dimensions of cognitive control more relevant than actions taken in the presence. With the progress in the field of Next Generation Sequencing driving testing in ordinary clinical settings, comes the obligation to ensure that providing patients with genetic risk information leads to patient benefit. In complex interventions such as clinical genetic services there are other relevant outcomes to consider besides health gain, especially for genetic diagnoses where no treatment is available. Empowerment may be such an outcome.

The chapter that focus on this aspect from clinical practical perspective is provided by Marion McAllister (Cardiff). She starts with an empirically founded critique of using risk perception as a measure of outcome from genetic counselling and proposes an alternative of 'risk recall' to empower patients. Her proposed model of empowerment in genetic services comprises five dimensions valued by patients and clinicians: cognitive, behavioural and decisional control, emotional regulation and hope. For practical guidance, a valid, reliable and responsive patient-reported outcome measure (the so-called Genetic Counselling Outcome Scale (GCOS-24) was developed. It allows to detect a significant improvement in patient empowerment levels following genetic counselling and testing.

The next chapter, provided by Karim Raza (Birmingham), Marie Falahee (Birmingham), Mats Hansson (Uppsala) discusses the relationship between medicalization and empowerment in a special setting of common, complex disease, namely rheumatoid arthritis (RA). For this disease, genetic profiling with identification of biomarkers is estimated to enable prediction and facilitate early treatment as well as preventive interventions for those individuals who are at increased risk but without a clinically manifest arthritis. The authors discuss the pros and cons of risk prediction. On the one hand, providing information about personal risk may be seen as an aspect of medicalization, where healthy individuals are labelled as 'at risk' and therefore responsible for taking preventive actions. On the other hand, it may provide helpful information that will enable autonomous health-related decision-making. Moreover, there is a risk that preventive testing and treatment may be requested and used more by some groups in society than others. Such a development risks increasing differences in health between different groups. As the example of RA also indicates, early intervention is associated with improved outcomes, but predictive approaches need to be accompanied by appropriate educational and psychological support. Therefore the authors suggest a concrete clinical framework of empowerment to support the development of personalized medicine and predictive testing in order to evade the pitfalls of medicalization and loss of autonomous control.

3. 'Rethinking professional responsibility' is the joint theme of the next two chapters. Both reflect as ethicist from empirical insights into the limits of existing concepts. Ulrik Kihlbom (Uppsala) critically revises the concept of 'understanding' risk by relying on insights from an interview study with Swedish haematologists on risk communication with leukaemia patients. Patients diagnosed with leukaemia may face a hard decision between a palliative approach or transplantation with allogeneic stem cells. Here, he examines in depth what the professional means when they claim that most patients cannot 'really understand' what the side-effects amount to. This observation functions as a vehicle for a more general discussion regarding the general aims of getting patients to understand risks. First, it will be discussed what kind of understanding of risks that the haematologists are after. Second, their motivation for trying so hard to get the patients to 'really understand', and, third, one consideration is put forward as a reason why their bleak attitude might be correct.

In her chapter on health risk communication and changing professional responsibilities, Silke Schicktanz (Göttingen) discusses from an ethical point of view how professional responsibilities exceed simplistic notions of giving honest and correct information when it comes to health risk information. Here, she enfolds the complexity of what professional responsibility entails, including how not only the patient, but also family members, are the object of responsibility, and that norms and consequences vary depending on the power held by the professionals. Furthermore, the epistemic dimension is important, how far behaviour really determines future health outcomes and, finally, we need to distinguish responsibility ascription whether they are backwards or future-oriented. Schicktanz discusses three cases of 'new genetics and health risks', such as breast cancer risk, biomarker testing in oncology and dementia prevention based on so-called risk prediction. Each case indicates that ethical norms related to professional responsibility can vary depending on the potential consequences that are implied for patients if/if not they follow the risk advice.

4. Social and policy dimensions. The first social implication is that not only the patient but also the family matters. Sabine Wöhlke (Göttingen), Marie Falahee (Birmingham), and Katharina Beier (Göttingen) jointly analysed how predictive health medicine recontextualizes medical decision-making. The chapter discusses how genetic decision-making creates a tension between individual accounts of autonomy (which dominate bioethical discourses) and family members' genetic interrelatedness, which transcends the logic of individual decision-making. The mere availability of genetic information can impose severe psychosocial and economic burdens on families. Ethical reflections on genetic testing must take these familial interdependencies into account. To this end, this chapter will discuss different aspects of family responsibility and how these present challenges to decision-making about genetic risk information. The theoretical reflections are illustrated by empirical data relating to predictive testing for particular diseases. In this way we will argue that intra-familial problems particularly arise a) when genetic knowledge challenges established familial practices of reciprocity and solidarity; b) when acquired practical experiences with specific diseases within the family influence genetic knowledge; and c) when genetic technology challenges prevalent understandings of family or even human nature in general.

Genomics has also left the clinical space and entered public domain: offering tests via the internet or making them so easily accessible that it requires new considerations of the rather traditional understanding of the genetic counselling in a doctor–patient relationship. Public communication tools, new media and non-professional organizations (e.g. patient organizations) become important factors impacting guidance and orientation for the public. Also, health policy is not bound any more to national regulation, but often lies across national borders. Therefore, international guidelines and exchange about good practice increases in importance. How this is challenged by Direct To Consumer Personal Genomic Testing (DTC PGT) is shown in the comparative focus group study in Germany, Israel, the Netherlands and the UK, provided by a pan-European author team lead by Manuel Schaper with Aviad Raz, Marie Falahee, Karim Raza, Danielle Timmermans, Elisa Garcia Gonzalez, Silke Schicktanz and Sabine Wöhlke as co-authors. In this chapter, they explore lay perspectives in Germany, Israel, the Netherlands and the UK with regard to acceptance of

DTC PGT, its perceived utility, its benefits and risks and its regulation and how these aspects relate to trust. They analysed 16 focus groups (participants $n=99$) using qualitative content analysis and identified variations in acceptance and concerns regarding DTC PGT based on aspects relevant to building trust. Privacy and data protection, including the possibility of discrimination, were a near-universal concern in all four countries. Regulation of the market was suggested to be necessary to protect consumers from possible negative consequences. Such comparative studies provide important insights into where transnational policy making should progress.

Public consumption of DTC PGT presupposes sufficient level of public genomic literacy, which may be lacking. The next contribution focuses on the importance of improving general public literacy on basic genomic concepts. A team of psychologists, Serena Oliveri, Renato Mainetti, Ilaria Cutica, Alessandra Gorini, Gabriella Pravettoni (Milano), start with the important observation that genomic testing for common diseases is increasingly available to the general public without any medical mediator. The public's level of understanding of basic biology and mathematics (including probability theory, statistics and risk) is still very poor. However, the knowledge of scientific terminology is essential for proper interpretation of genetic risk information. The authors therefore analyse how people elaborate genetic knowledge for disease predisposition: cognitive and psychological aspects, beliefs, perceptions about disease controllability and decisions. By arguing for the importance of improving general public literacy in genetics and 'empowering' people to make informed decisions, also through policy initiatives, they suggest the application of new technologies (mobile applications, web sites, etc.), and serious games in particular, to foster genetic knowledge and education.

The last contribution of this volume considers the use of a relatively new tool in processing health and genomic information in medical decision-making, viz., artificial intelligence (AI). Machine learning algorithms may be very useful in processing complex and large amounts of information in decision-making. This chapter looks especially at the potential use of AI at the regulatory level of decision-making. It draws on conceptual as well as empirical research to analyse the implications of ongoing development of AI as part of regulatory decision-making in medicines.

The intended readership of this volume is interdisciplinary. Students and researchers in theoretical and clinical medical ethics, medical sociology, medical psychology, as well as professionals in clinical genetics and lawyers, may be addressed by different chapters. Furthermore, this volume is very international and transcends traditional foci where ethics and law often produce a consistent, but also narrow thought system. In the volume, the studies refer to different countries, with a focus on wider Europe.

Part I

New visions of 'genetics and genomics' as integrative medicine

2 Genomic medicine in 2025–2030

Martina C. Cornel
and GertJan van Ommen

Medical genetics as a medical speciality

Medical genetics developed some decades ago as a medical specialty for questions on paediatric patients and consequences for future pregnancies. Specialized physicians and clinical geneticists in many countries offered genetic services in clinical genetic centres, where laboratory testing was also organized as well as support from genetic counsellors. When a child was born with a developmental problem or multiple congenital anomalies, the etiological diagnosis in this child helped to make decisions on their treatment and typically it would assist in determining the recurrence risk of following pregnancies. Then prenatal diagnosis became available, and later also preimplantation genetic diagnosis in combination with in vitro fertilization. Subsequently oncogenetics and cardiogenetics led to an increasing number of referrals to clinical genetic centres. Inherited tumour syndromes were recognized and it was understood that the identified genetic background of sudden death at a young age could have consequences for relatives. DNA testing for inherited BReast CAncer (*BRCA*) and colorectal cancer (Lynch syndrome) became possible, as well as familial hypercholesterolemia, inherited arrhythmias and cardiomyopathies. For oncogenetics and cardiogenetics, preventive strategies and early detection in mutation carriers were the main drivers of increasing patient numbers: knowing that a person carries a *BRCA* mutation for instance allows carriers to opt for prophylactic surgeries and/or surveillance at a younger age than is offered in the population screening program for breast cancer.

More involvement of non-genetic experts

Now that more and more people may profit from a DNA-test, ordering one is increasingly done by non-genetic experts, although local situations differ. Paediatricians order tests for specific developmental problems, oncologists order a DNA-test if the outcome may affect the choice of their treatment. As more and more tests for common complex disorders are becoming available, they will move to mainstream medicine, simply because of the resources. This increasing role of non-genetic experts in genetic services is referred to as 'mainstreaming'. The proportion of physicians using DNA-testing in their everyday work is increasing, while the number of clinical geneticists and genetic counsellors trained to communicate DNA-test results is limited. Specific physicians are trained to work with specific DNA-tests themselves. In some countries mainstreaming has been pioneered in research projects, like ordering *BRCA*-tests in ovarian cancer patients (George et al. 2016). Geneticists devote their time to relatives at 50 per cent risk of carrying a mutation, while the oncologist gets fast test results to determine the choice of therapy. In the course of this development, some questioned to what extent day-to-day primary care can benefit from detailed genomic knowledge (Hayward et al. 2017). Much of this discussion took place in the UK, where the Department of Health funded the 100,000 Genomes Project, intending to integrate genomics medicine in the NHS. The project is expected to accelerate the pace of moving genomics knowledge to everyday health care. This led to diagnoses of rare diseases, some of which were treatable once the diagnosis was known. It also led to unforeseen new applications of sequencing in health care, such as to identify different strains of tuberculosis, meaning patients can be treated with precisely the right medication more quickly (www.genomicsengland.co.uk).

New ethical issues

The definitions and goals of 'genetic counselling' have changed over time and differ between countries. Clinical genetics started to develop for parents with severely affected infants. These couples potentially have to make difficult reproductive decisions. If a diagnosis of, e.g., an autosomal recessive condition was made in a child, this would imply a 25 per cent recurrence risk for each future child within the family. Options for the family might include:

1. refrain from having (more) (biologically own) children
2. use prenatal diagnosis, and in the case of an affected foetus, termination of pregnancy
3. use donor gametes.

All these raise their own ethical issues. An important goal of genetic services was 'helping genetic counselling clients cope with and adapt to genetic information' as well as to support 'client-informed decision-making' (Biesecker 2001). In some cultures, health care professionals stressed the possibilities of using genetic knowledge to prevent birth defects and genetic disorders, and wanted to inform clients so that they would use genetic knowledge 'to make rational or logical reproductive choices' (Biesecker 2001). In other cultures, this goal of prevention was not considered acceptable, and instead an important goal of genetic counselling was to help couples 'choose a course of action which seems to them appropriate in view of their risk, their family goals, and their ethical and religious standards and act in accordance with that decision' (Fraser 1974). The latter stems from a document of a subcommittee of the American Society of Human Genetics in 1974. In the subsequent decades there was a notable shift from the 'prevention paradigm' to informing clients in a non-directive way and supporting them to make informed decisions.

With the increasing possibilities in oncogenetics and cardiogenetics, there are possibilities of prevention that are less ethically sensitive. The call to inform people on the variants in their genome that might be relevant for prevention is gaining ground. The 100,000 Genomes Project, for instance, proposes to return to participants information on bowel cancer predisposition, breast and ovarian cancer predisposition, other cancer predisposition, familial hypercholesterolaemia and carrier status (Genomics England 2019). No matter what the reason to participate in the project, all participants can opt to receive these additional or secondary findings. The gene variants at stake 'may cause an increased risk of certain genetic diseases. These diseases can often can be prevented or reduced by NHS treatment.' As for carrier information: 'Some gene changes (or variants) do not affect the individual that has them, but may affect future children.' The NHS is recruiting patients with cancer or rare diseases, but these additional or secondary findings are not related to the condition 'that led them to take part' (Genomics England 2019).

A different way to express a similar change in ethical perspective would be the shift from the 'right-not-to-know' to a 'duty to inform'. For the paradigmatic case of Huntington disease many have discussed whether it would be advantageous for people to be informed of their carrier status of Huntington disease before they develop symptoms (Mastromauro et al. 1987). This autosomal dominant neurodegenerative disorder leads to cognitive decline, together with chorea and personality change. Life expectancy is reduced. A pre-symptomatic diagnosis may allow people to make certain informed choices (e.g. to have a family or not; to make a career or travel), but many have also warned that this knowledge may increase anxiety and suicide risk, disturb family relationships and impact the freedom of choice of any children. It was thus proposed that 'The professionals should not promote nor advocate pre-symptomatic DNA-testing' (Robins Wahlin 2007). The more diseases become treatable or preventable, the more a 'duty to inform' can be defended. Currently, the hope of therapies for Huntington is growing (Bashir 2019). This may lead to an increasing uptake of pre-symptomatic DNA-testing in Huntington families, and simultaneously to more support for a right to be informed. In 2025–2030 the right-to-know will play a central role in genomic medicine.

One gene, a few genes, many genes, comprehensive risk models

When it became possible to do a DNA-test to investigate the cause of a monogenic disease, for instance breast cancer, initially testing was done gene by gene. If no mutation was found in *BRCA1*, subsequently *BRCA2* was investigated. In the last decade, sequencing technology made it possible to investigate many more genes simultaneously. This led to the development of gene panels: several high-risk genes for a certain condition could be investigated at the same time and with relatively limited resources. Not only *BRCA1* and *BRCA2*, but also *ATM, CDH1, CHEK2, PALB2* and *TP53*, which are all associated with an increased cancer risk (LaDuca et al. 2019). A gene panel test can simultaneously analyse these high-risk genes and many medium-risk genes.

A different development is the polygenic risk score (PRS), building on a multitude of low-risk genes (Khera et al. 2018; Mavaddat et al. 2019). Each of these gene variants, Single Nucleotide Polymorphisms (SNPs), confers a very small increase of the breast cancer risk, but hundreds of SNPs are combined in a score that may help to stratify risk. Chatterjee et al. (2016) proposed to use these polygenic risk scores to determine for each woman when to start breast cancer screening. The average women in the UK had a 2.4 per cent breast cancer risk over a ten-year period at the recommended starting age for screening in 2016. Chatterjee proposed to determine individual risks and identify the optimal age to start surveillance accordingly. Thus, the highest risk quintile would start mammographic screening before 40 years of age, while the lowest risk quintile would not reach the 2.4 per cent threshold before 80 years of age, and not need surveillance.

Combining all relevant risk factors (high-risk genes, polygenic risk score, mammographic density, age, lifestyle, hormonal, reproductive, anthropomorphic factors) in one comprehensive breast cancer prediction model (BOADICEA) will enable optimal risk stratification (Lee et al. 2019). The implementation of such a comprehensive model has started. Next steps include adapting the algorithm to local situations and obtaining legal approval (CE IVD). In 2025–2030 every woman can fill out a family history questionnaire at 25 years of age and have a PRS, including high-risk, medium-risk and low-risk variants, after which she is stratified for breast cancer screening at the age and interval personalized to her situation. Similar personalized surveillance programs for colon cancer and prostate cancer will be developed.

Genomics in oncology: guiding treatment and identifying relatives at risk

When sending a tumour sample to a pathologist, increasingly the analysis includes molecular testing that can lead to the diagnosis of a germline mutation. Whether or not a tumour will respond to therapy based on a specific monoclonal antibody for instance is determined by the activity of the relevant receptor. An early example is Her2Neu overexpression determining whether Herceptin® may be effective. If testing for a gene mutation (e.g. *BRCA*) is performed because it can determine treatment response, the test for a somatic mutation will also find germline mutations. Thus, the pathologist may find symptoms suggestive of inherited breast and ovarian cancer (*BRCA* mutations, *BRCA*-ness) or Lynch syndrome (e.g. MSI/IHC: absent MSH6) (Byrum et al. 2019; Snowsill et al. 2017). After finding such a symptom, he may consider asking for referral of the patient or family. Pathologists also perform genomic sequencing in tumour tissue. When doing so, they identify both somatic and germline mutations (e.g. somatic *KRAS* or *BRAF* mutations as well as germline *BRCA* mutations). By 2025–2030 both treatment of cancer patients and information for relatives will build on the tumour testing by pathologists.

Shift to prevention

Genomic information can identify people at a high risk of developing cardiovascular disorders or cancer before they develop the disease. In some cases, interventions are available to reduce their risk of having, e.g., myocardial infarction or colon cancer. Where this combination of a high predictive ability and interventions are available, genomics may help to prevent symptoms. With an increasing proportion of elderly people in Western populations, this is urgently needed (P4 medicine: Predictive, Preventive, Personalized and Participatory). Already some decades ago in the Netherlands, a cascade screening program for familial hypercholesterolemia (FH) was implemented, because statins contributed so many years to the life expectancy of these patients. FH is often the consequence of a heterozygous mutation in one of three genes. Nowadays polygenic risk scores (PRS) using information on hundreds of genes are becoming available for implementation in health care. Initially, their added value was heavily debated, as they were derived from too small population sample sets and typically failed to be validated. However, a 2018 landmark paper by Kathiresan and co-workers using 100,000 samples from the first phase of the UK biobank and validated in a second 100,000 phase, showed the predictive ability of PRS for risk stratification in seven major common disorders: coronary artery disease, atrial fibrillation, type 2 diabetes, inflammatory bowel disease, and breast cancer (Khera et al. 2018). It was soon thereafter realized that even a million variants would easily fit on a SNP array costing less than $50, which would make this a viable test for people at significantly higher risk than the population average. Thus, costlier follow-up, diagnosis and prevention could be focused on a much smaller at-risk population, substantially contributing to cost-effective disease prevention. For example, at the time, screening for Familial Hypercholesterolemia, a monogenic cause of high-cholesterol-mediated myocardial infarction, was implemented in some countries given its high frequency of 1:200–1:500 and the availability of preventive statin treatment. In developing this screening, the risk boundary for such a protocol was typically considered as greater than 3–4x population risk. However, PRS screening for a similar high-risk coronary artery disease yielded a more than tenfold greater population (Khera et al. 2018). Not detecting these, while it would take a $50 chip to do so, with the same preventive option of statins, was increasingly argued to be suboptimal prevention. The same would hold true for several forms of cancer which typically are diagnosed too late for curative intervention, like breast/ovarium and colon cancer. In 2018–2022, considering that the basis to develop these common disease PRS chips was large enough, with fully sequenced sample sets with phenotypic information, and that their precision would be highest when they were available for the background populations in each country, a pan-European initiative was started, called the 'Million Genomes' project (European '1+ Million Genomes' Initiative 2019), aimed at making such datasets available for research if present, and generating them if not. This ultimately allowed the reduction of the number of variants required for robust prediction by 2025–2030 to a few thousand, driving down the cost of the PRS chips in participating countries to under $10.

The citizen as keeper of his own data

Simultaneous with the increase in the possibilities of 'big data' there is a shift occurring in the patient–doctor relationship (Brill et al. 2019). Health information technologies make it possible for patients and healthy citizens to monitor their health using devices, read their electronic health record and analyse their own whole genome sequence. Citizens could upload their heartbeat and blood pressure data before their visit to a physician for a health check. They can use eHealth tools to monitor their chronic conditions (Lancaster et al. 2018). By reporting back symptoms and medication use, the dosage can be adapted after online communication. Thus, chronic disease patients could visit their health care provider only when symptoms deteriorated, and online communication would not suffice. Citizens and patients thus increasingly take responsibility for their own health.

For genomic sequence data, patients and research participants could access and store their data on their own device, and analyse when, where and what they consider relevant. There is growing support for the view that people should have access to their raw (uninterpreted) data upon request (Thorogood et al. 2018). In 2019 health care systems, there is, however, a limited understanding of the clinical validity and utility of genome variants. In 2025–2030 if a patient goes to his physician taking his genome on a laptop, physicians will know how to deal with this. Should a patient need a prescription of 5-fluorouracil, the oncologist will find it very practical to look into the genomic data for the *DPYD*-genotype, and prescribe a personalized dosage, to avoid adverse effects. If this prescription is given to a cancer patient needing pain control, also the *CYP2D6*-genotype can be studied before starting codeine or tramadol treatment, to optimize pain management. If the patient has a relapse of ovarian cancer, the data of *BRCA1&2* can be analysed for germline mutations to decide if PARP-inhibitors are indicated. If the data is available, these analyses can be done instantly, like opening a window to a specific part of the sequence data. No additional time and resources will be needed.

Citizens might find other analyses much more interesting, such as information on ancestry (the percentage of Neanderthal DNA, the percentage of Scandinavian DNA) or the likelihood to have the hair colour that they have (Janssens 2019). These recreational uses of DNA-data in general help to better understand genomics. There are also potential harms, such as finding out about a high risk of a serious condition (ApoE genotype and Alzheimer's disease) or about biological kinship (paternity issues). (Messner 2011; Moray et al. 2017). By the 2025–2030 recreational use of genomic data will be ubiquitous, but children as well as medical issues will be protected by law.

Different countries and different organizations might find different solutions to help citizens to keep their own data. As a general principle, individuals have a legal right to access their personal data held by governmental bodies and commercial entities and patients have a right to access their health record (Thorogood et al. 2018). With the increasing possibilities of using genomic data in risk prediction, citizens can take more responsibility for their own health by using this data. Health care thus needs to be prepared for the healthy person visiting a physician because he has a genetic variant and asking for subsequent care. Care pathways for people carrying an oncogenetic variant or cardiogenetic variant must be clear. It must also be clear which variants to interpret as relevant versus not significant. The challenges of the interpretation of Variants of Uncertain Significance (VUSs) will transcend into general medicine.

Genomics based treatment

Understanding the etiologic mechanism behind diseases makes it possible to treat very precisely. This is true for many fields of medicine: cancer, infectious diseases, common complex diseases, rare monogenic conditions. Gene therapies that were developed around the year 2000 led to vector-related leukaemia in some patients, after which it took a while to find other approaches to gene therapies (Cavazzana et al. 2016). In recent years, however, we have seen approval of several very promising somatic gene therapies (Cornel et al. 2019; Al-Zaidy and Mendell 2019; Ikawa et al. 2019). Challenges include the regulatory process both in terms of access to the market and reimbursement. Since many of the gene therapies are targeting small patient groups with clinical heterogeneity, the traditional methodology in evidence-based medicine, using Randomized Controlled Trials (RCTs) involving large numbers of patients, is not suitable. New designs to prove the efficacy and effectiveness of treatment for (conditional) approval have been proposed, including *N* = 1 trials and Bayesian approaches (Aartsma-Rus 2011; Abrahamyan et al. 2016). In 2025–2030 evidence synthesis practices have moved beyond RCTs with increased recognition of the importance of pragmatic and patient-oriented evidence (Tingley et al. 2018).

With the success of new gene therapies on the one hand, affordability becomes a challenge: high investments may lead to high prices. In May 2019, Novartis® announced that the price of a therapy for spinal muscular atrophy (SMA) would be $2 million for one infusion, the most expensive drug ever (Cohen 2019; Kuchler 2019). There were arguments to show that this price was cost-effective, since current therapies cost more over a ten-year period (Cohen 2019). The new drug can be a one-time, life-saving treatment. However, this is a comparative and market-based approach and it does not address earlier investments from public resources; earlier research was largely funded by grants from charities and government agencies. Since non-profit organizations and NIH invested in the development of the drug, a 'reasonable price' is discussed (Love 2019). In 2025–2030, over 200 gene therapies are available for prices below 20,000€ per Quality Adjusted Life Year.

For CRISPR, clinical trials have started, but so far neither the procedure for approval by regulatory bodies nor the price of treatment is clear (Cornel et al. 2019). CRISPR-based repair of specific mutations could be done in centres experienced in stem cell transplantation and CRISPR: drawing blood, isolating stem cells, modifying these cells *ex vivo*, and transfusing them back to the individual patient. The dispensing may resemble magistral drug preparation, a model that may circumvent many of the financial challenges (Schellekens et al. 2017). Since CRISPR is targeting many different mutations, a flexible bedside solution is likely to be in place in a few years' time. Unlike in the pharmaceutical industry, magistral preparation takes place in the hospital pharmacy overseen by a physician for the patients under their care. Bedside development and magistral drug production would be an affordable, safe and flexible alternative for precision treatment of patients. At least in theory, a 'Bio-Nespresso'-machine 'allows the pharmacist to produce medicine for only one individual patient which does not have to comply with the regulations commercially produced medicine have to' (Bisnajak 2019). If production-specific supplies are available in the pharmacy or CRISPR facility, a tailored solution for every patient could be prepared at a 'small-scale manufacturing unit for personalized medicine production' (Bisnajak 2019). In discussions about CRISPR therapies, some state that it is 'safe, cheap and effective'. So far, however, gene therapies are quite expensive, because the procedures for market authorization and reimbursement require investments, and commercial investors require return of investment. In 2025–2030 safe and effective gene therapy is available and affordable for large groups of patients.

Take home messages:

- Pre-symptomatic genomic information increasingly allows for early detection and prevention of disease.
- Implementation of genomic medicine will involve all sectors of health care.
- Personalized precision medicine holds promise for the treatment of many genetic conditions, including cancer.
- Magistral production of individualized $N = 1$ gene therapies may be an acceptable bypass of randomized clinical trials for rare diseases.

References

Aartsma-Rus A. (2011), The risks of therapeutic misconception and individual patient (n = 1) 'trials' in rare diseases such as Duchenne dystrophy. *Neuromuscular Disord.*, 21(1):13–5. doi: 10.1016/j.nmd.2010.09.012.

Abrahamyan L, Feldman BM, Tomlinson G, Faughnan ME, Johnson SR, Diamond IR, Gupta S. (2016), Alternative designs for clinical trials in rare diseases. *Am J Med Genet C Semin Med Genet.*, 172:313–31. doi: 10.1002/ajmg.c.31533.

Al-Zaidy SA, Mendell JR. (2919), From Clinical Trials to Clinical Practice: Practical Considerations for Gene Replacement Therapy in SMA Type 1. *Pediatr Neurol.*, 13(18): 31163–9. doi: 10.1016/j.pediatrneurol.2019.06.007.

Bashir H. (2019), Emerging therapies in Huntington's disease. *Expert Rev Neurother.* 17:1–13. doi: 10.1080/14737175.2019.1631161.

Biesecker BB. (2001), Goals of genetic counselling. *Clin Genet*, 60: 323–30.

Bisnajak A. (2019), The Bio-Nespresso Project: The design of a small-scale manufacturing unit for personalized medicine production. Available from https://repository.tudelft.nl/islandora/object/uuid%3A2bc01b5e-cbb6-4465-b172-fba54e1bf368 Accessed Sept 19, 2019.

Brill SB, Moss KO, Prater L. (2019), Transformation of the Doctor-Patient Relationship: Big Data, Accountable Care, and Predictive Health Analytics. HEC Forum. Jun 17. doi: 10.1007/s10730-019-09377-5.

Byrum AK, Vindigni A, Mosammaparast N. (2019), Defining and Modulating 'BRCAness'. *Trends Cell Biol*, 29:740–51. doi: 10.1016/j.tcb.2019.06.005.

Cavazzana M, Six E, Lagresle-Peyrou C, André-Schmutz I, Hacein-Bey-Abina S. (2016), Gene Therapy for X-Linked Severe Combined Immunodeficiency: Where Do We Stand? *Hum Gene Ther*, 27(2):108–16. doi: 10.1089/hum.2015.137.

Chatterjee N, Shi J, García-Closas M. (2016), Developing and evaluating polygenic risk prediction models for stratified disease prevention. *Nat Rev Genet*, 17: 392–406. doi: 10.1038/nrg.2016.27.

Cohen J. (2019), At Over $2 Million Zolgensma is the World's Most Expensive Therapy, Yet Relatively Cost-Effective. Available from: www.forbes.com/sites/joshuacohen/2019/06/05/at-over-2-million-zolgensma-is-the-worlds-most-expensive-therapy-yet-relatively-cost-effective/#42575e3245f5 Accessed Sept 19, 2019.

Cornel MC, Howard HC, Lim D, Bonham VL, Wartiovaara K. (2019), Moving towards a cure in genetics: what is needed to bring somatic gene therapy to the clinic? *Eur J Hum Genet*, 27:484–87. doi: 10.1038/s41431-018-0309-x.

European '1+ Million Genomes' Initiative. (2019), Available from: https://ec.europa.eu/digital-single-market/en/european-1-million-genomes-initiative. Accessed Oct 7, 2019.

Fraser FC. (1974), Genetic counseling, *Am J Human Genet*, 26:636–61. Available from www.ncbi.nlm.nih.gov/pmc/articles/PMC1762720/pdf/ajhg00444-0104.pdf Accessed Sept 19, 2019.

Genomics England. (2019), What can participants find out? Available from: www.genomicsengland.co.uk/information-for-participants/findings/ Accessed Sept 19, 2019.

George A, Riddell D, Seal S, Talukdar S, Mahamdallie S, Ruark E, Cloke V, Slade I, Kemp Z, Gore M, Strydom A, Banerjee S, Hanson H, Rahman N. (2016), Implementing rapid, robust, cost-effective, patient-centred, routine genetic testing in ovarian cancer patients. *Sci Rep*. Jul 13, 6:29506. doi: 10.1038/srep29506.

Hayward J, Bishop M, Rafi I, Davison V. (2017), Genomics in routine clinical care: what does this mean for primary care? *Br J Gen Pract*, 67(655):58–9. doi: 10.3399/bjgp17X688945.

Ikawa Y, Miccio A, Magrin E, Kwiatkowski JL, Rivella S, Cavazzana M. (2019), Gene therapy of hemoglobinopathies: progress and future challenges. *Hum Mol Genet*, 19. pii: ddz172. doi: 10.1093/hmg/ddz172.

Janssens C. (2019), Cecile, your genetics make you unlikely to have red hair. Twitter Feb 20, 2019. Available from https://twitter.com/cecilejanssens/status/1098313274024701954 Accessed Sept 19, 2019.

Khera AV, Chaffin M, Aragam KG, Haas ME, Roselli C, Choi SH, et al. (2018), Genome-wide polygenic scores for common diseases identify individuals with risk equivalent to monogenic mutations. *Nat Genet*, 50(9):1219–24. doi: 10.1038/s41588-018-0183-z.

Kuchler H. Novartis wins approval for world's most expensive drug. Available from: www.ft.com/content/10086870-7e50-11e9-81d2-f785092ab560 Accessed August 7, 2019.

LaDuca H, Polley EC, Yussuf A, Hoang L, Gutierrez S, Hart SN, et al. (2019) A clinical guide to hereditary cancer panel testing: evaluation of gene-specific cancer associations and sensitivity of genetic testing criteria in a cohort of 165,000 high-risk patients. *Genet Med*, 13. doi: 10.1038/s41436-019-0633-8.

Lancaster K, Abuzour A, Khaira M, Mathers A, Chan A, Bui V, et al. (2018), The Use and Effects of Electronic Health Tools for Patient Self-Monitoring and Reporting of Outcomes Following Medication Use: Systematic Review. *J Med Internet Res*, 20:e294. doi: 10.2196/jmir.9284.

Lee A, Mavaddat N, Wilcox AN, Cunningham AP, Carver T, Hartley S, et al. (2019), BOADICEA: a comprehensive breast cancer risk prediction model incorporating genetic and nongenetic risk factors. *Genet Med*, 21(8):1708–18. doi: 10.1038/s41436-018-0406-9.

Love J. (2019), Why didn't nonprofits and the NIH require 'reasonable' pricing for Zolgensma? That may happen in France. Available from: www.statnews.com/2019/09/18/zolgensma-reasonable-pricing-france/ Accessed Sept 19, 2019.

Mastromauro C, Myers RH, Berkman B. (1987), Attitudes toward presymptomatic testing in Huntington disease. *Am J Med Genet*, 26:271–82. doi: 10.1002/ajmg.1320260205.

Mavaddat N, Michailidou K, Dennis J, Lush M, Fachal L, Lee A, et al. (2019), Polygenic risk scores for prediction of breast cancer and breast cancer subtypes. *Am J Hum Genet*, 104(1):21–34. doi: 10.1016/j.ajhg.2018.11.002.

Messner DA. (2011), Informed Choice in Direct-to-Consumer Genetic Testing for Alzheimer and Other Diseases: Lessons from Two Cases. *New Genet Soc*, 30:59–72. doi: 10.1080/14636778.2011.552300

Moray N, Pink KE, Borry P, Larmuseau MH. (2017), Paternity testing under the cloak of recreational genetics. *Eur J Hum Genet*, 25(6):768–70. doi: 10.1038/ejhg.2017.31.

Robins Wahlin TB. (2007), To know or not to know: a review of behaviour and suicidal ideation in preclinical Huntington's disease. *Patient Educ Couns*, 65:279–87.

Schellekens H, Aldosari M, Talsma H, Mastrobattista E. (2017), Making individualized drugs a reality. *Nat Biotechnol*, 35:507–13. doi: 10.1038/nbt.3888.

Snowsill T, Coelho H, Huxley N, Jones-Hughes T, Briscoe S, Frayling IM, Hyde C. (2017), Molecular testing for Lynch syndrome in people with colorectal cancer: systematic reviews and economic evaluation. *Health Technol Assess*, 21:1–238. doi: 10.3310/hta21510.

Thorogood A, Bobe J, Prainsack B, Middleton A, Scott E, Nelson S, et al. (2018), Participant Values Task Team of the Global Alliance for Genomics and Health. APPLaUD: access for patients and participants to individual level uninterpreted genomic data. *Hum Genomics*, 12:7. doi: 10.1186/s40246-018-0139-5.

Tingley K, Coyle D, Graham ID, Sikora L, Chakraborty P, Wilson K, et al. (2018), Using a meta-narrative literature review and focus groups with key stakeholders to identify perceived challenges and solutions for generating robust evidence on the effectiveness of treatments for rare diseases. Orphanet. *J Rare Dis*, 13:104. doi: 10.1186/s13023-018-0851-1.

Part II

Risk empowerment

3 Genomic risk perception and implications for patient outcomes from genetic counselling

Marion McAllister

The model of empowerment comprises five dimensions valued by patients and clinicians: cognitive, behavioural and decisional control, emotional regulation and hope. Findings were used to develop the Genetic Counselling Outcome Scale (GCOS-24), which was shown to be a valid, reliable and responsive patient-reported outcome measure, able to detect significant improvement in patient empowerment levels following genetic counselling and testing. Finally, I describe application of GCOS-24 to service evaluation and continuous quality improvement in the All Wales Medical Genetics Service.

One of the important goals of genetic (genomic) counselling is the education of patients and families regarding the chances of disease occurrence or recurrence (Resta et al., 2006). For this reason, one active area of research in genetic counselling has been to assess the accuracy with which patients report or recall risk figures before and after genetic counselling, often reporting this as 'risk perception'. Improvements in 'accuracy' of perceived risk after genetic counselling have been reported, but not consistently, and there is a tendency for risk to be overestimated after genetic counselling (Butow et al., 2003; Meiser and Halliday, 2002; Sivell et al., 2008; Smerecnik et al., 2009; Wang et al., 2004).

However, we have known for a long time that patients of clinical genetics services often struggle to 'rationalize' the risk figures provided in genetic counselling. For example, some patients have been reported to translate genetic risks into 'binary' form – the event (e.g. a pregnancy affected by the family condition) either will or will not happen. In this way, patients may simplify probabilistic information to shift their focus to the implications of being at risk and the potential impact of what might or might not occur (Lippman-Hand and Fraser, 1979; Parsons and Atkinson, 1982).

Why might this be? Are genetic counsellors ineffective at communicating genetic risks? Are genetic counselling patients not clever enough to understand risk figures? This question was discussed by Jehannine Austin (2010), who argued that whilst genetic counsellors may use the term 'risk' as a synonym for 'numeric probability', genetic counselling patients may develop a composite understanding of risk that combines the numeric probability provided in genetic counselling with their perceived severity of the family condition as well as contextual and other factors to arrive at a 'sense' of their own risk. Austin recommended that genetic counsellors avoid use of the term 'risk' with patients and focus instead on numeric probabilities when communicating 'risk' figures. Austin further argues that using 'risk recall' as a measure of 'risk perception accuracy' is erroneous because numeric probabilities only become meaningful to the patient when integrated with their own context and their perceived severity of the family condition. The tendency to overestimate the numeric probabilities provided in genetic counselling is likely to, at least in part, reflect the patient's perceived severity of the family condition. Furthermore, a patient's 'sense' of risk may change over time with the unfolding of events relating to the family condition. Austin argued that for this reason, recall of risk figures (or 'risk perception') is not a good measure of outcome from genetic counselling.

In the social sciences, of course, the idea that 'objective' scientifically calculated risks are correct and 'subjective' (perceived) risks are incorrect has been challenged (Pidgeon et al., 1992; Douglas and Wildavsky, 1982). It has been argued that we now live in a 'risk culture' (Giddens, 1991; Beck, 1992), despite the fact that the term 'risk' may have lost its associations with probability. Instead, we focus more on the magnitude of the danger than on the probability of the outcome (Douglas, 1992). This is in keeping with Austin's assertion that the severity of the family condition has a significant impact on genetic counselling patients' risk perception, and the tendency for overestimation of perceived risk over the genetic probability of occurrence or recurrence of the family condition.

In my own qualitative research, conducted in the 1990s following patients through pre-symptomatic genetic testing for Lynch Syndrome, an inherited cancer condition, I found that lay ideas about how physical characteristics are inherited in families could also have a significant influence on genetic risk perception. Participants often talked about how family members 'take after' one parent more than another, often the same sex parent, suggesting that participants made assumptions about physical and other characteristics being inherited in sets, not as individual features. These ideas appeared to draw upon lay models of inheritance that are common in our culture (Richards, 1996a; 1996b) to make inferences about co-inheritance of Lynch Syndrome with some other characteristic, e.g. facial features, build or sex (McAllister 2002; 2003). Richards (1996c) has described how we are socialised from an early age with ideas about how physical and other characteristics are inherited, noting how we welcome a new child into our family by discussing who she 'takes after'. Because these ideas are present from our earliest years, they become implicit in our thinking about inheritance, and could impede us from taking on 'new' (to non-geneticists) explanations of how inheritance works. Indeed, some participants in my study could reiterate to me in the interviews the genetic model of inheritance provided to them in genetic counselling, and it was clear that they understood that this model predicts a 50 per cent risk to the children of a family member affected by Lynch Syndrome. However, some participants would then go on to directly contradict this, justified by referring to an assumption that because they did or did not share certain other characteristics with their affected parent, that they would or would not inherit Lynch Syndrome too.

To further complicate this picture, some participants appeared to be making use of lay models of inheritance to help them to cope with the upcoming result of their pre-symptomatic genetic test for Lynch Syndrome. Interestingly, multiple individuals from the same families were interviewed, and the lay models described by participants were specific to the individuals themselves, rather than being shared in their family. The models used drew upon both existing lay models of inheritance that are shared in our culture and also upon information specific to their own family. Some participants, whose preferred coping strategy was to 'prepare for the worst', appeared to pay increasing attention to negative aspects of family history, and to select a cognitive 'rule' (lay model of inheritance) from a set of possible 'rules' about the inheritance of physical and other characteristics; one that predicted themselves to be a carrier of the family 'cancer gene'. These 'intensely engaged' participants became convinced that they would be found to carry the family 'cancer gene' and experienced feelings of fear and anxiety at the prospect of bad news related to a pre-symptomatic genetic test result. They appeared to use these patterns of thinking to move toward greater certainty in the face of the uncertainty of their 50 per cent risk status as they awaited the result of their genetic test. This also meant that in the event of a bad news test result, these participants had already done 'the work of worry' and appeared to adapt well to their result. Some other 'partially engaged' participants, who were happier to adopt a 'wait and see' approach, protecting themselves from the anxiety of anticipating bad news, were somewhat less well-prepared for a bad news test result. These participants struggled a little more to adapt to a bad news test result, and this appeared to be partly because they had not done the anticipatory 'work of worry'.

It appeared that participants may have been using lay models of inheritance as a type of 'fast and frugal heuristic'. Fast and frugal heuristics (Gigerenzer et al., 1999; Goldstein and Gigerenzer, 2002; Gigerenzer et al., 2007), whilst they can be sources of cognitive errors and biases, can also be adaptive in the sense that they can provide a decision-maker with adaptive cognitive strategies. The work of Gigerenzer and colleagues has shown that in some situations, heuristic reasoning can be associated with better choices. Furthermore, in the context of genetic risk, predictions are impossible, and uncertainty can only be quantified with provision of genetic probability information. From the patient's perspective, however, it may be that the event either will or won't happen and perhaps any cognitive strategy that enables the patient to cope with that uncertainty might be considered adaptive.

There was also some evidence that participants' risk perception could change with the unfolding of time and Lynch Syndrome-related events in their family. For example, some participants described the intensity of their convictions that they would develop cancer at the same age as their affected parent did, and then when they passed through this 'risky age' without developing cancer, their assumption was that they had 'escaped'. It was also apparent that the quality and quantity of participants' experiences of cancer-related suffering and death amongst relatives, and the closeness of their relationships with affected family members could also have a significant influence on their risk perception and the intensity of their engagement with their cancer risk.

All this seems to confirm that genetic counselling patients' perception of genetic risk can be a complicated matter, and that it can be significantly influenced by their own context and the perceived severity of the family condition, supporting the arguments made by Austin. So, if recall of risk figures (or 'risk perception') is not a good measure of genetic counselling effectiveness, what is? A wide range of other outcomes have been measured and there is evidence that genetic counselling can result in improvements in patients' knowledge, perceived personal control and health behaviours, and psychological distress. Furthermore, there is evidence that most patients are very satisfied after genetic counselling (Payne et al., 2008; Madlensky et al., 2017).

As part of a programme of research conducted between 2003 and 2010 at Nowgen in Manchester, one of five 'genetics knowledge parks' funded by the UK Department of Health, a systematic review of patient-reported outcome measures (PROMs) used to evaluate clinical genetics services was completed (Payne et al., 2008). The review demonstrated that (i) the psychometric quality of measures available at that time was poor and (ii) the diversity of measures resulted in confusion about what PROs were appropriate. The views of genetic counselling patients and providers were sought about what they value as outcomes from genetic counselling. A series of focus groups and interviews were conducted with key stakeholders (McAllister et al., 2008; 2011a) and transcripts were analysed using grounded theory. The emerging theoretical framework summarized patient benefits from genetic counselling and testing using the term 'empowerment' which comprised the following five dimensions:

1 Decisional control: knowledge about options that are available for the family condition/risk management and feeling able to make informed decisions between those options.
2 Cognitive control: sense-making or feeling that one has the best available explanation for the condition in the family, understanding the genetic risks to oneself and other relatives and future generations and knowing what health and social support is available.
3 Behavioural control: feeling able to use health and social care systems to manage the family condition.

4 Hope: for a fulfilling family life for oneself and one's relatives.
5 Emotional regulation: feeling able to cope with the family condition, and able to manage one's feelings about the family condition and risks.

Cognitive, decisional and behavioural control together comprise the construct perceived personal control, which had previously been operationalized as a PROM for evaluation of genetic counselling (Berkenstadt et al., 1999; Payne et al., 2008). The addition of hope and emotional regulation broadens the empowerment construct to better encapsulate the patient benefits from genetic counselling. Perceptions of control are known to be important in chronic disease, with perceived control associated with less mood disturbance (Affleck et al., 1987; Scharloo et al., 1998).

Could empowerment be a useful PRO for evaluating genetic counselling and related services? At that time, no single measure was available that captured all the dimensions of empowerment, so it seemed the logical next step to develop a good quality PROM to capture this. The Genetic Counselling Outcome Scale (GCOS-24) was designed to capture empowerment and comprises 24 questions with seven Likert-style response categories (McAllister et al., 2011b).

GCOS-24 was tested using psychometric methods which demonstrated that GCOS-24 has good internal consistency (Cronbach's $\alpha = 0.87$) and test re-test reliability (intra-class correlation = 0.86). Hypothesis testing confirmed that responses to GCOS-24 correlate positively with perceived personal control, health locus of control and authenticity, and negatively with anxiety, as expected, demonstrating construct validity. Assessment of sensitivity to change (responsiveness) of GCOS-24 demonstrated that GCOS-24 can detect statistically significant improvements in GCOS-24 scores following genetic counselling with a medium-to-large effect size (Cohen's $d = 0.7$) (McAllister et al., 2011b). The Minimum Clinically Important Difference for GCOS-24 was established to be a score increase of 10.3 points following genetic counselling (Thomas and McAllister, 2019). This means that the smallest change in GCOS-24 scores following genetic counselling that is meaningful for and important to patients is 10.3 points, contributing to interpretability of GCOS-24 scores.

GCOS-24 has been used to evaluate genetic counselling services in the UK and Canada, where patients have consistently been shown to have significantly improved GCOS-24 scores following clinic attendance (Costal Tirado et al., 2017; Inglis et al., 2014; McAllister, 2016). In Cardiff, post-clinic GCOS-24 scores were found to correlate significantly with patient satisfaction, and to be useful for service quality improvement (Costal Tirado et al., 2017). These studies demonstrated for the first time that clinical genetic counselling services can deliver measurable patient benefits. GCOS-24 has potential for routine evaluation of genetic counselling services and could offer PRO data useful to decision-makers, e.g. commissioners and funders of those services.

GCOS-24 has been translated and adapted for use into Denmark (Diness et al., 2017), Spain (Munoz-Caballo et al., 2018) and the Netherlands (Voorwinden et al., 2019). Brazilian Portuguese and Japanese translations are in preparation. In 2019, a short six-item version of the GCOS-24 (The Genomics Outcome Scale (or GOS)) was developed with five Likert-style response categories (Grant et al., 2019). Using methods derived from Item Response Theory, GOS has been demonstrated to maintain the ability of GCOS-24 to capture empowerment, whilst providing a less burdensome scale for respondents.

References

Affleck G, Tennen H, Pfeiffer C, Fifield J. 1987. Appraisals of control and predictability in adapting to a chronic disease. *J Personal Soc Psychol*;53(2):273–9.

Austin J. 2010. Re-conceptualizing risk in genetic counseling: implications for clinical practice. *J Genet* Couns;19(3):228–34.

Beck U. 1992. *Risk Society: Towards a New Modernity*. New Delhi: Sage.

Berkenstadt M, Shiloh S, Barkai G, Bat-Miriam-Katznelson M, Goldman B. 1999. Perceived personal control (PPC): a new concept in measuring outcome of genetic counseling. *Am J Med Genet*;82: 53–9.

Butow PN, Lobb EA, Meiser B, Barratt A, Tucker KM. 2003. Psychological outcomes and risk perception after genetic testing and counselling in breast cancer: a systematic review. *Med J Aust*;178(2):77–81.

Costal Tirado A, McDermott AM, Thomas C, Ferrick D, Harris J, Edwards A, McAllister M. 2017. Using patient-reported outcome measures for quality improvement in clinical genetics: an exploratory study. *J Genet Couns* 26(5): 1017–28.

Diness BR, Overbeck G, Hortshoj, TD, Hammer TB, Timshel S, Sorensen E, McAllister M. 2017. Translation and adaptation of the Genetic Counselling Outcome Scale (GCOS-24) for use in Denmark. *J Genet Couns*;26(5):1080–9.

Douglas M. 1992. *Risk and Blame: Essays in Cultural Theory*. London: Routledge.

Douglas M, Wildavsky A. 1982. *Risk and Culture*. Berkeley, CA: University of California Press, p. 221.

Giddens A. 1991. *Modernity and Self-identity: Self and Society in the Late Modern Age*. Cambridge: Polity Press.

Gigerenzer G, Todd PM, and the ABC Research Group. 1999. *Simple heuristics that make us smart*. New York: Oxford University Press.

Gigerenzer G, Gaissmaier W, Kurz-Milcke E, Schwartz LM, Woloshin S. 2007. Helping doctors and patients make sense of health statistics. *Psychological Science in the Public Interest*;8(2):53–96.

Goldstein DG, Gigerenzer G. 2002. Models of ecological rationality: The recognition heuristic. *Psychological Review*;109:75–90.

Grant PE, Pampaka M, Payne K, Clarke A, McAllister M. 2019. Developing a short-form of the Genetic Counselling Outcome Scale: The Genomics Outcome Scale. *Eur J Med Genet*;62(5):324–34.

Inglis A, Koehn D, McGillivray B, Stewart SE, Austin J. 2014. Evaluating a unique, specialist psychiatric genetic counseling clinic: uptake and impact. *Clinical Genetics*;EPub 22 May. doi: 10.1111/cge.12415.

Lippman-Hand A, Fraser FC. 1979. Genetic counseling–the postcounseling period: I. Parents' perceptions of uncertainty. *Am J Med Genet*;4:51–71.

Madlensky L, Trepanier AM, Cragun D, Lerner B, Shannon KM, Zierhut H. 2017. A Rapid Systematic Review of Outcomes Studies in Genetic Counseling. *J Genet Counsel*;26(3):361–78.

McAllister M. 2002. Predictive genetic testing and beyond: A theory of engagement. *J Hlth Psychol*;7:491–508.

McAllister M. 2003. Personal theories of inheritance, coping strategies, risk perception and engagement in Hereditary Non-polyposis Colon Cancer families offered genetic testing. *Clinical Genetics*;64:179–89.

McAllister M. 2016. Genomics and patient empowerment. In: Kumar D, Chadwick R., eds. *Genomics and Society*. London: Elsevier Inc.

McAllister M, Dunn G, Todd C. 2011a. Empowerment: Qualitative Underpinning of a New Patient Reported Outcome for Clinical Genetics Services. *Eur J Hum Genet*;19:125–30.

McAllister M, Payne K, Nicholls S, MacLoed R, Donnai D, Davies L. 2008. Patient empowerment in clinical genetics services. *J Health Psychol*;13:887–97.

McAllister M, Wood A, Dunn G, Shiloh S, Todd C. 2011b. The Genetic Counseling Outcome Scale: a new patient-reported outcome measure for clinical genetics services. *Clinical Genetics*;79:413–24.

Meiser B, Halliday J. 2002. What is the impact of genetic counselling in women at increased risk of developing hereditary breast cancer? A meta-analytic review. *Soc Sci Med*;54(10):1463–70.

Munoz-Caballo P, Garcia-Minaur S, Espinell-Vallejo ME, Fernandez-Franco L, Stephens A, Santos-Simarro F, Lapunzina-Badia P, McAllister M. 2018. Translation and Cross-Cultural Adaptation with Preliminary Validation of GCOS-24 for Use in Spain. *J Genet Couns*;27(3):732–43.

Parsons E, Atkinson P. 1992. Lay constructions of genetic risk. *Soc Hlth Illness*;14:438–55.

Payne K, Nicholls S, McAllister M, MacLeod R, Donnai D, Davies L. 2008. Outcome Measurement in Clinical Genetics Services: A systematic review of validated measures. *Value in Health*;11(3):497–508.

Pidgeon NF, Hood C, Jones D, Turner BA. 1992. *Risk perception*. Risk: Analysis, Perception and Management, The Royal Society, London; 89–134.

Resta R, Bowles Biesecker B, Bennett RL, Blum S, Estabrooks Hanh S, Strecker MN, Williams JL. 2006. A New Definition of Genetic Counseling: National Society of Genetic Counselors' Task Force Report. *Journal of Genetic Counseling*; 15: 77–83.

Richards MPM. 1996a. Lay and professional knowledge of genetics and inheritance. *Public Und Sci*;5(3):217–30.

Richards MPM. 1996b. Lay knowledge of inheritance and genetic risk. A review and hypothesis. *Health Care Anal*;4:1861–4.

Richards MPR. 1996c. Families. In: Marteau T, Richards M, eds. *The Troubled Helix: Social and Psychological Implications of the New Human Genetics*. Cambridge: Cambridge University Press, 249–73.

Scharloo M, Kaptein AA, Weinman J, Hazes JM, Willems LNA, Bergman W. Rooijmans HGM. 1998. Illness perceptions, coping and functioning in patients with rheumatoid arthritis, chronic obstructive pulmonary disease and psoriasis. *J Psychosomatic Res*;44:573–85.

Sivell S, Elwyn G, Gaff CL, Clarke AJ, Iredale R, Shaw C, Dundon J, Thornton H, Edwards A. 2008. How risk is perceived, constructed and interpreted by clients in clinical genetics, and the effects on decision making: systematic review. *J Genet Counsel*;17(1):30–63.

Smerecnik CM, Mesters I, Verweij E, de Vries NK, de Vries H. 2009. A systematic review of the impact of genetic counseling on risk perception accuracy. *J Genet Counsel*;18(3):217–28.

Thomas C, McAllister M. 2019. Establishing the Minimum Clinically Important Difference for the Genetic Counseling Outcome Scale (GCOS-24). *J Genet Couns* [EPub 30 July 2019].

Voorwinden JS, Plantinga M, Krijnen W, Ausems M, Knoers N, Velthuizen M, Birnie E, Lucassen AM, van Langen IM, Ranchor AV. 2019. A validated PROM in genetic counselling: the psychometric properties of the Dutch version of the Genetic Counselling Outcome Scale. *Eur J Hum Genet*;27(5):681–90.

Wang C, Gonzalez R, Merajver SD. 2004. Assessment of genetic testing and related counseling services: current research and future directions. *Soc Sci Med*;58:1427–42.

4 Genomic and biological risk profiling

From medicalization to empowerment

Mats G. Hansson, Karim Raza and Marie Falahee

Introduction

The dominant current trend in medicine is towards more precise targeting of individual characteristics related to genotype, biological markers and environmental factors that are decisive for diagnosis, treatment and prevention of disease. This development has been called *personalized* or precision medicine. Risk profiling is an integral part of the stratified therapeutic and preventive medical intervention in accordance with individual characteristics and needs. For complex disorders, including some types of heart disease, arthritis and cancer, the resulting phenotype is due to a variation in many genes, each contributing a small amount of genetic susceptibility, acting in concert with multiple environmental or epigenetic factors. Some environmental risk factors might be modifiable, e.g. smoking, nutrition, exercise, alcohol intake, and others less so, e.g. environmental pollution or psychosocial stress.

For rheumatoid arthritis (RA), risk profiling on the basis of predictive biomarkers and early symptoms could facilitate early treatment as well as preventive interventions for those individuals who are at increased risk of developing RA but are without a clinically manifest arthritis. The European League Against Rheumatism (EULAR) recommendations identify five such groups that are appropriate for prospective trials: individuals having: (i) genetic risk factors for RA; (ii) environmental risk factors for RA; (iii) systemic autoimmunity associated with RA; (iv) symptoms without clinical arthritis (arthralgia); or (v) unclassified arthritis (Gerlag et al., 2012). Within the *Mind the Risk* project we have studied perceptions of risk and predictive testing held by a range of stakeholders, including patients and at risk groups (Stack et al., 2016; Falahee et al., 2017; Simons et al., 2018; Mosor et al., 2019; Falahee et al., 2018, Bayliss et al., 2018).

We discuss some of the results here and reflect on the *pros* and *cons* of risk prediction. On the one hand, providing information about personal risk may be seen as an aspect of medicalization where healthy individuals are labelled as 'at risk' and are therefore responsible for taking preventive actions. On the other hand, it may provide helpful information that will enable autonomous health-related decision-making and an increased possibility for preventive measures.

Genetic and biological risk profiling as an example of medicalization

Medicalization has been conceived of as a socially problematic expansion of medical science and the expansion of the authority of medical doctors into social sectors where problems were earlier solely defined and solutions proposed based on social and psychological theories. This is how it was conceptualized by Irving K Zola in 1972 (Zola, 1972). Medicine was described as increasingly becoming an institution of social control through medicalizing much of daily living, by making medicine and the labels healthy and ill *relevant* to an ever increasing part of human existence. Medical scientists and doctors became experts in societal fields far outside their competences. Socio-economic factors contributing to illness and health may be played down and the role of an individual's behaviour in illness assigned greater prominence. Pregnancy was an early identified case of a natural process that has become increasingly conceptualized as a medical condition with increasing requirements for the monitoring and intervention of women's lifestyle behaviours.

As knowledge of how environmental and behavioural factors affect illness and health increases, there is a growing emphasis on identifying patients before they develop any symptoms. As stated by Zola: 'The physician must not only seek out his clientele but once found must often convince them that they must do something *now* and perhaps at a time when the potential patient feels well or not especially troubled' (Zola, 1972, p. 493). The central idea of primary prevention is to reach out to individuals while they are still healthy in order to prevent illness. Accordingly, medical practitioners now regularly ask their clients about smoking, alcohol consumption or physical exercise, and prescribe both pharmacological and behavioural preventive interventions.

For Zola and the tradition around him, medicalization was an inappropriate and abusive exercise of medical authority, intruding into private lives and expanding social control to areas where it was not justified. Part of the claim was that 'medicine has focused on the diseases of the rich and the established – cancer, heart disease, stroke – and ignored the diseases of the poor, such as malnutrition and still high infant mortality' (Zola, 1972, p. 489). Interestingly, however, the development in genetics points to a significant difference in this regard since; while environmental factors like bad housing and unhealthy lifestyles may follow socio-economic gradients, many genetic diseases seem to hit in a more random way, quite often with no explanation available with regard to environmental factors. Cancer, arthritis, chromosomal aberrations and rare disorders due to random mutations affect both rich and poor.

Recent research in epigenetics may shift the balance, putting even more emphasis on how environmental factors affect the transcription of DNA. There is growing evidence of the impact of non-genetic variables on gene expression, e.g. nutrition, maternal care, lifestyle behaviour, psychosocial stress, adversity and neglect in early life. Molecular epigenetic mechanisms in the cell explain how the exposure to environmental factors influences the phenotypic outcome and variability both between individuals and within an individual at different times (Meaney, 2001; Vågerö et al., 2018). However, many genetic diseases will, to an important extent, remain a random event, affecting individuals in a 'democratic' way, giving little room for external social control. For some genetic diseases, social factors and lifestyle factors may be a triggering event.

For Marcel Verweij, the main issue in medicalization is not about social control but about the loss of autonomy of individuals to be able to lead a life as they choose. The increased focus on prevention of disease carries with it also an increased risk of, what he calls, iatrogenic risks of prevention due to false positive or negative test results and 'persons may sometimes be subjected to invasive diagnostic procedures or even mutilating surgery' (Verweij, 1999, p. 95). Probabilistic estimations of disease risk due to environmental and lifestyle factors are made at a population level, and do not reveal any definitive guidance for the individual, or rule in/out a disease. For example, Verweij stated: 'Smoking is by far the most important cause of lung cancer; still it is rather difficult to prove that a person would have stayed healthy if (s)he had not smoked cigarettes' (op. cit., p. 100). This is reflected by a recurring theme across our studies of predictive and preventive approaches to RA; that negative viewpoints were associated with appreciation of the probabilistic nature of risk information, and willingness to take preventive action was linked to the assumption that this would rule out future development of RA:

> If someone could confirm to me, 100%, that if I stopped smoking and had a healthier diet I could stop myself from developing RA, then I would consider it – I would try to stop smoking and live a healthier life.
>
> (First degree relative of RA patient)

However, as Verweij also has pointed out, there are indeed positive effects of an increased focus on medical and biological factors affecting health. Infant mortality has undoubtedly decreased due to an increased focus on preventive factors related to the medicalization of pregnancy and birth. Lung cancer has dropped significantly due to campaigns against smoking and cardiovascular diseases are better controlled by use of statins as a preventive intervention. For cardiovascular diseases it has been shown that 90 per cent of the total risk is related to modifiable factors, e.g. smoking, physical exercise, psychosocial stress (Yusuf et al., 2004). Furthermore, as argued by Jonathan Scholl: 'Medicalization can even involve increasing one's feeling of self-control, such as with birth control and medicalized pregnancies' (Scholl, 2017).

For Verweij, the major problem concerning medicalization related to prevention is related to personal autonomy. He claims that 'the practice of preventive medicine publicly confirms and sustains views of life in which health is most important, and that, consequently, it becomes more difficult for individuals to live a life and embrace a view of life in which other values than health are most important. In a liberal-democratic society, it is considered morally problematic if public institutions hinder persons developing and practicing their own views of life' (Verweij, 1999, p. 105). There is a loss of personal autonomy that, according to Verweij, is a major moral problem. In such a society, 'it is not easy to develop and maintain a carefree, happy-go-lucky attitude towards one's own health' (ibid.). However, health-promoting information may balance other socio-cultural influences that shape people's views, e.g. advertising by tobacco companies; widespread availability of cheap, convenient, unhealthy foods etc. It could equally be argued that from this viewpoint preventive approaches enhance, rather than hinder, autonomy by facilitating informed viewpoints. Furthermore, even if preventive medicine may sustain a view that health is very important or 'most important', it doesn't follow that public institutions hinder persons having their way of life. The individual remains in both principle and in practice free to follow the advice or to focus on other values. Except for certain medical practices, such as transplantation where there is a shortage of available organs and one needs to set priorities regarding the likely effect of a transplantation due to the lifestyle of the recipient, individuals are at liberty to make life choices associated with substantial health risks (e.g. rock climbing, boxing). When the costs of an unhealthy lifestyle will affect others, access to limited resources may be prioritized, e.g. a liver transplant for an alcoholic wanting to stay on his 'happy-go-lucky' lifestyle should be given to someone with an attitude that is more likely to imply an effective use of the scarce resource. We need public institutions in order to make and administer this kind of balancing in transparent and democratic ways.

In a society with unlimited resources, the latitude for not attending to one's own health may be large, but in practice, and in most societies, the economic space for choosing risky lifestyles or for not trying to prevent ill-health is limited. Accordingly, vascular surgery may be associated with requirements to stop smoking. Giving one individual a share in health care may have the consequence that someone else will not have a share. Protection of personal autonomy is important but is limited when this results in negative consequences for others.

In order to be able to practice autonomy, one needs also to have information, e.g. regarding the effects of unhealthy 'happy-go-lucky' lifestyles for one's health or views of life. We wish to live our lives according to our own cherished values of what makes a good, enjoyable or meaningful life. However, an individual's 'values' are also socially/culturally constructed, and shaped by many factors, e.g. media, governments through legislation, family, commercial interests etc., and we would also like to have the opportunity to change a course of life and even change our salient values. Values do not in themselves constitute sufficient arguments for selecting a course of life. One would also like to preserve the option to sometimes change one's opinions after having acquired more information about the facts, or having perceived the kind of value conflicts which arise, when some value which one esteems is achieved. One perhaps discovers values which had passed unnoticed and undesirable consequences which had not been anticipated, but which one would like to acknowledge. We tend to agree with George Henrik von Wright's idea that informed preferences should be taken more seriously than the uninformed preferences we actually happen to have at the moment: 'To come into possession of, or experience some X which we wish, increases our welfare provided that we would wish this X if we were informed about the causal relations and consequences which hold both for the totality of which X is part and the totality where not-X is included instead of X' (von Wright, 1997). It doesn't follow that such information should be imposed on people but it should be made available, e.g. in the format of public health information, leaving the moral and decisional authority to make choices at the hands of the individual, as long as others' choices of life are not affected.

This account of the importance of informed preferences and the possibility of choosing your own way of life in accordance with such preferences is in accord with how the concept of health is sometimes understood. Verweij suggests that the growing practice of preventive medicine may actually be detrimental to the confidence that people may have in their health. Risk information may lead both to better planning for the individual, but it may equally well lead to increased worry and anxiety, something that should be appropriately addressed in preventive strategies/risk counselling. From a public health perspective and within the context of acknowledging the interest of others for enjoyment of scarce health care resources people at risk should maybe be worried enough to change a risky lifestyle behavior. However, that depends both on how accurate the information is and if the information is actionable. Lennart Nordenfelt has elaborated on the concept of health along these lines (Nordenfelt, 2000). His notion of health is defined thus: a person is healthy, if and only if he or she is in a bodily and mental state so that he or she can realize their vital goals. This, in turn, requires balanced reflection about one's goals and, accordingly, some information to support this.

Health risk information based on gene sequencing or the prevalence of different biomarkers is a kind of information for which questions about medicalization may be particularly pertinent. Risk profiling as part of medical practice fits very well within Zola's remark about the very idea of primary prevention as providing health-related information to individuals in order to convince them to 'do something *now* and perhaps at a time when the potential patient feels well or not especially troubled' (op. cit., p. 493). Undoubtedly, it would be a negative type of medicalization if the information is not actionable and only leads to increased worry, or even wrong actions taken due to inaccurate information based on tests with too low sensitivity or specificity. However, when risk profiling provides opportunities for very early diagnosis with associated preventive treatment or effective changes in lifestyle it would be part of a positive type of medicalization. Potentially, it would even lead to the empowerment of individuals and families receiving this kind of information. However, this depends partly on how 'empowerment' is defined.

Genetic and biological risk profiling as an example of empowerment

Eric Juengst et al. have suggested that genetic profiling as a means to personalize medicine needs to supported by 'evidence of the extent to which genetic information brings actionable intelligence for patients. If the information given to patients has no effect on their actions, then the promise of patient empowerment is empty' (Juengst, 2012). Indeed, the literature so far has been unable to provide any significant amount of this kind of evidence. Theresa Marteau's Cochrane Review from 2010 (updated 2016) showed no evidence of behavioural change due to receiving genetic information (Marteau et al., 2010; Hollands et al., 2016). They concluded that this kind of risk communication has little or no effect on smoking and physical activity. It may have a small effect on self-reported diet and on intentions to change behavior but does not overall lead to any behavioural changes. Claims that genomic risk profiling motivates people to change their lifestyle and thereby actively engage in prevention seems unfounded and, accordingly, one may conclude that promises about empowerment are unfounded. However, the picture is more complex, and it depends partly on how 'empowerment' is defined. Sparks et al. have provided evidence that providing personalized risk (not just genetic) information to first degree relatives of RA patients had a positive effect on health intentions and behaviours (Sparks et al., 2018). It is also well known that mutation carriers tend to experience an increase in psychological distress shortly after obtaining test results but their distress return to pre-testing levels over time. So genetic testing may not cause long-term psychological harm – but this is not the same as achieving a positive change. One needs more in order to claim empowerment.

The move towards patient-centered care has resulted in an increased focus on patient empowerment. Tools for activating, engaging, informing and communicating with patients should presumably be evaluated from the perspective of the benefits they bring to patients and in terms of patients' satisfaction with their decision, or decisional regret. Hibbard et al. in Oregon developed a patient activation measure (PAM) (Hibbard et al., 2004). This is a self-reported questionnaire that assess respondents' views on knowledge, skills and confidence for self-management of their health or chronic condition. Several similar instruments have been used to capture patient activation and patient engagement. A problem in this field of research is that there is no consensus on how to define and distinguish between the different concepts of patient empowerment, engagement and activation. A recent scoping review by Risling et al. to explore the concept and use of patient empowerment within the electronic health context demonstrated that in total 19 empirical publications (out of 1387) had used related measures, but with a significant variety in their operationalization of the concept, with concepts such as patient activation, empowerment and engagement being used interchangeably (Risling et al., 2017). There is accordingly a great need for conceptual clarification in order to understand what measures should be used to evaluate the provision of information related to disease risk and opportunities for risk reduction.

Marion McAllister and her team developed the concept of empowerment as an outcome of genetic risk counselling based on a comprehensive review of psychometric instruments measuring related constructs (Bravo et al., 2015). (See Chapter 3 in this volume for a fuller account.) Her account of empowerment will put Juengst's claim that genetic risk information doesn't lead to empowerment into perspective. Empowerment is conceptualized as a multi-dimensional construct that is not only related to capacity to change behaviour. Empowerment comprises a broad range of domains (health status, psychological distress, coping, decision-making, family functioning, knowledge and risk perception, perceived personal control, quality of life, self-esteem and spiritual well-being). McAllister et al. grouped these into five dimensions: i) decisional control, ii) cognitive control, iii) behavioural control, iv) hope and, v) emotional regulation on the basis of their review of related outcome measures (Payne et al., 2008). This study reviewed 67 validated Patient Reported Outcome Measurements and assessed them regarding their psychometric properties. Of these, 30 were developed specifically for genetic risk information, 37 were generic.

McAllister et al. developed a measure of empowerment incorporating these five dimensions that was validated for the assessment of empowerment as an outcome from the provision of genetic information in the UK National Clinical Genetic Services (McAllister et al., 2011).

The five dimensions were:

1 Cognitive control, i.e. patients having access to understandable and updated information about treatment and research processes.
2 Decisional control, i.e. patients knowing what options are available regarding treatment and for participating with their data in research

and feeling able to make informed decisions between available options that sustain their values and concerns.

3 Behavioural control, i.e. patients feeling able to use health and social care systems effectively as well as research to reduce harm and/or improve their lives and those of their relatives and future descendants.
4 Hope, i.e. patients having hope for a fulfilling family life, for oneself, relatives and future descendants.
5 Emotional regulation, i.e. patients feeling able to cope with their condition (in the family), feeling able to manage feelings of distress and anxiety.

For the remainder of this chapter we present examples from a transcript of a previously unpublished focus group conducted as part of the evaluation of a novel informational resource for individuals at risk of developing rheumatoid arthritis. We focus on quotations that illustrate either aspects of medicalization or each of the five dimensions of McAllister's conceptualization of empowerment.

Evaluation of risk information for individuals at risk of developing rheumatoid arthritis

Background

Two female patients (both white British, aged 33 and 44) with clinically suspect arthralgia (joint pain at risk of progression to rheumatoid arthritis) were recruited in an early arthritis clinic at Sandwell and West Birmingham Hospitals NHS Trust, UK to take part in a focus group. Before the focus group, participants were asked to read an informational resource entitled 'Am I likely to develop rheumatoid arthritis? A guide for people with joint symptoms', which was developed by an international multidisciplinary team of researchers, clinicians and patient research partners who were taking part in the EU funded project EuroTEAM ('Towards Early Arthritis Management'; grant agreement FP7-HEALTH-F2-2012-305549). The objective of the focus group was to discuss positive and negative aspects of the informational resource as part of a multi-stakeholder evaluation exercise. Before taking part in the focus group, participants signed to indicate informed consent to participate. The duration of the focus group was 74 minutes. The focus group was facilitated by two researchers and was audio-recorded. The anonymized recording was transcribed by an independent professional transcription company.

Medicalization

The experience of having been identified as a person at risk of developing RA was described as like being in 'No Man's Land', and difficult to articulate to others. They described a lack of information for those in this situation, and having previously had access only to information designed for those with established RA. The informational resource that was the topic of the focus group, which was directed specifically to support those at risk was viewed as a welcome development to address their concerns.

P1: When you're at this early stage, the joint symptoms, you might feel a bit like, 'What am I?' You're in No Man's Land. They say it to you like it's a good thing, don't they? 'Don't worry; you've not got rheumatoid arthritis.' You think, 'Oh, but then at least I'd have a label.' Like you say, it's hard to know what to say to people.

P2: Yeah, and it's hard to know what to think yourself. That's why this leaflet is quite good because it does explain that part of it.

Being at risk was described as 'a new version of healthy', though one participant wondered whether they could be required to state their 'at risk' status in legal documents and job applications, despite feeling healthy.

> I don't know whether that will affect my employability because at the moment, I'm just the same as I was before. I don't think there's anything... I don't have any symptoms. I feel completely normal but I have had these markers in my blood. Should I be stating them and if I didn't, would I be in trouble?

Some concern was raised about the idea that people are increasingly made to feel responsible for maintaining their health, and the use of language that reinforces this was critically highlighted:

> In medicine at the moment, there's a lot of responsibility on people for the idea which I don't really like always, especially things to do with weight. I even think here about – 'making sure you are not overweight' – I would have rather that was phrased as something like – 'Try to maintain a healthy weight' or something.

Empowerment

1 Cognitive control:

The informational resource under discussion during the focus group contained information about RA, RA risk factors, opportunities for risk reduction, and addressed the potential for both positive and negative consequences of learning about one's risk status. Resources for further information and support were also provided. As such the resource directly addressed the cognitive control dimension of McAllister's empowerment concept and was valued by the focus group participants who highlighted that previously they had only encountered information that addressed those with established RA, rather than those at risk of developing RA:

> I thought it was a good leaflet. I think if somebody would have given me this when I went to see the GP, I would have found it helpful because it explains rheumatoid arthritis but also gives lots of caveats that you might not develop it. It's just making you aware that there's a higher risk that you will; whereas, I was given some information about rheumatoid arthritis but not about pre-rheumatoid arthritis which I think are two different things. I found it, on the whole, helpful at putting things in perspective in a way that answered a lot of my questions.

2 Decisional control:

Participants described how the information presented in the informational resource was important to inform decision-making. For example, in relation to decision-making about whether to seek medical help for early symptoms of RA:

> It's valuable in the respect that they might decide not to do something at the moment but it is pointing out that – 'Okay, don't do anything at the moment but make sure that you do come forward if you need to.' I think a lot of people might say, 'Yeah, I'll just see how it goes for a bit but I do know I'd have to go and see the GP'... I think it might be useful information.

Participants valued the balance between positive and negative aspects of risk information, and described how this could affect decision-making about whether to accept testing to predict RA progression or to inform an insurance company or employer about the one's risk status:

> It could really prejudice an employer, you could feel and I think if people understand that, that would potentially put them right off going for a test.

3 Behavioural control:

Positive viewpoints were expressed about risk-related information that facilitates positive health outcomes, for example by promoting lifestyle change to reduce risk:

> If you know that you can actually do something and try and change your lifestyle and just to be aware that if something happens to you, go and check it out.

The value of balanced information was reiterated in this context, particularly in supporting decision-making around if and when to seek support from health professionals:

> There is a lot of information and it is quite balanced. It could make people ask about it to their GP which I don't think is a bad thing.

4 Emotional regulation:

Participants described the emotional challenges of being at risk of RA progression:

> I think it took me a long time to come to terms with the fact that I am probably quite likely to develop it and the symptoms will get worse.

They described the information resource as providing reassurance. For example, by dispelling misperceptions about RA and the likely outcomes associated with RA:

> Those are the things that I found quite reassuring to read because it did make me realize that it's not just older people. There are younger people than myself who've suffered with it and found ways to overcome it and they're suffering in far more extreme ways than I have. I think to know that, again, there are other people of that age, is reassuring at that stage when you're first being told about it.
>
> I just immediately had images of being deformed or being… disabled in some way until I was able to read more information of case studies of people being given treatment and living active lives.

5 Hope:

Both focus group participants described the risk information resource as a source of hope. Hope that they might not develop RA, and that there are positive interventions that could reduce their risk and effectively manage RA if they were to develop it.

> That's why I found the leaflet quite uplifting because it does say that there are good drugs that can do things, there is hope and the trials are mentioned in there, which I think is really good, and you can get involved in.
>
> Even though the rheumatoid arthritis information I saw did sort of put it in perspective, you didn't have anything actually in your hands saying, 'You might get it. You might not and there's hope.'

Discussion

Though there were some examples of medicalization, and negative aspects of being identified as an individual at risk, participants in this focus group valued the risk information resource and welcomed information that directly addressed those in a specific (high-) risk group, rather than disease related information designed for those with established RA.

There were illustrations of each of the five dimensions of McAllister et al.'s concept of empowerment in this focus group discussion. In particular participants highlighted the need for balanced information, addressing both positive and negative aspects of risk information, and this balance perhaps provides a useful strategy to empower informed decision-making and counter claims of medicalization associated with predictive and preventive strategies.

Although information about disease risk has potential to cause anxiety, and this was highlighted by participants across our qualitative studies within the *Mind the Risk* project, the informational resource under discussion in this focus group was valued by participants as a source of reassurance and hope. A randomized control trial of a personalized risk education tool for first-degree relatives of RA patients has demonstrated that risk information lowered concern about RA and RA risk (Marshall et al., 2018), and that participants overestimated their risk of developing RA prior to the educational intervention. Our studies of patients and individuals at risk have emphasized that predictive and preventive approaches to RA need to be accompanied by appropriate provision of informational and psychological support, and that this is particularly important for individuals at high risk of developing RA, such as those with early symptoms (Mosor et al., 2019). The resource under evaluation in the focus group described here represents an important first step towards addressing this need.

The qualitative RA studies for *Mind the Risk* have informed the development of ongoing quantitative studies to assess and predict the preferences of those at risk of developing RA for predictive and preventive approaches. This information will facilitate the development of tailored, patient-centric strategies, addressing both positive and negative aspects of such approaches, and empower patients to make informed, supported decisions in relation to management of their risk status and well-being.

It should be observed that the concept of empowerment as outlined and discussed here has a very strong focus on the individual and the specific genetic counselling context, e.g. the consultations that take place at a hospital ward between a clinical geneticist/counsellor and a proband. This perspective needs indeed to be complemented with a focus on conditions needed for achieving social control that provides the means for the patient/proband to actually make a choice.

References

Bayliss K, Raza K, Simons G, Falahee M, Hansson M, Starling B, Stack RJ, (2018) Perceptions of predictive testing for those at risk of developing a chronic inflammatory disease: a meta-synthesis of qualitative studies, *Journal of Risk Research* 21(2):167–89.

Bravo P, Edwards A, Barrs PJ, Scholl I, Elwyn G, McAllister M, (2015) Conceptualising patient empowerment: a mixed methods study, *BMC Health Services Research* Jul 1;15:252. doi: 10.1186/s12913-015-0907-z.

Falahee M, Simons G, Buckley CD, Hansson M, Stack RJ, Raza K, (2017) Patients' Perceptions of Their Relatives' Risk of Developing Rheumatoid Arthritis and of the Potential for Risk Communication, Prediction, and Modulation, *Arthritis Care* Res Oct;69(10):1558–65.

Falahee M, Simons G, Raza K, Stack R, (2018) Healthcare professionals' perceptions of risk in the context of genetic testing for the prediction of chronic disease: a qualitative metasynthesis, *Journal of Risk Research* 21(2):129–66.

Gerlag DM, et al. (2012) EULAR Recommendations for Terminology and Research in Individuals at Risk of Rheumatoid Arthritis: Report from the Study Group for Risk Factors for Rheumatoid Arthritis, *Ann Rheum Dis* May;71(5):638–41.

Hibbard JH, Mahoney ER, Stockard J, Tusler M, (2004) Development and testing of a short form of the patient activation measure, *Health Services Research* 39(4):1005–26.

Hollands G, French DP, Griffin SJ, Prevost AT, Sutton S, King S, Marteau TM, (2016) The impact of communication genetic risks of disease on risk-reducing health behaviour: a systematic review with meta-analysis, *BMJ* 352:i1102. https://doi.org/10.1186/s41927-018-0038-3.

Juengst Eric T, Flatt Michael A, Settersten Jr Richard A, (2012) Personalized genomic medicine and the rhetoric of empowerment, *Hastings Center Report* 42(5):34–40.

Marshall AA, Zaccardelli A, Yu Z, et al. 2018. Effect of communicating personalized rheumatoid arthritis risk on concern for developing RA: A randomized controlled trial, *Patient Educ Couns*. 102(5):976–83. doi: 10.1016/j.pec.2018.12.011.

Marteau TM, French DP, Griffin SJ, et al., (2010) Effects of communicating DNA-based disease risk estimates on risk-reducing behaviours, *Cochrane Database Syst Rev*. Oct 6;(10):CD007275. doi: 10.1002/14651858.CD007275.pub2.

McAllister M, Wood AM, Dunn G, Shiloh S, Todd C, (2011) The Genetic Counseling Outcome Scale: a new patient-reported outcome measure for clinical genetic services, *Clinical Genetics* 79(5):413–24.

Meaney MJ, (2001) Maternal care, gene expression, and the transmission of individual differences in stress reactivity across generations. *Annual review of neuroscience* 24(1):1161–92.

Mosor E, Stoffer-Marx M, Steiner G, et al. (2019) I would never take preventive medication! Perspectives and information needs of people who underwent predictive tests for rheumatoid arthritis, *Arthritis Care and Res* https://doi.org/10.1002/acr.23841.

Nordenfelt L, (2000) *Action, ability and health – Essays in the philosophy of Action and Welfare*, Kluwer Academic Publishers, Dordrecht.

Payne K, Nicholls S, McAllister M, Macleod R, Donnai D, Davies LM, (2008) Outcome measurement in clinical genetics services: a systematic review of validated measures, *Value Health* 11(3):497–508.

Risling T, Maryinez J, Young J, Thorp-Froslie N, (2017) Evaluating patient empowerment in association with eHealth technology: Scoping Review, *Journal of Medical Internet Research* 9(9), e 329.

Scholl J, (2017) The muddle of medicalization: pathologizing or medicalizing?, *Bioethics* 38:265–78, quote on p. 273.

Simons G, Stack RJ, Stoffer-Marx M, et al., (2018) Perceptions of first-degree relatives of patients with rheumatoid arthritis about lifestyle modifications and pharmacological interventions to reduce the risk of rheumatoid arthritis development: a qualitative interview study, *BMC Rheumatology* 2:31.

Sparks JA, Iversen MD, Yu Z, et al., (2018) Disclosure of Personalized Rheumatoid Arthritis Risk Using Genetics, Biomarkers, and Lifestyle Factors to Motivate Health Behavior Improvements: A Randomized Controlled Trial, *Arthritis Care Res* 70(6):823–33.

Stack RJ, Stoffer M, Englbrecht M, et al., (2016) Perceptions of risk and predictive testing held by the first-degree relatives of patients with rheumatoid arthritis in England, Austria and Germany: a qualitative study. *BMJOpen* 6:e010555. doi: 10.1136/bmjopen-2015-010555.

Vågerö D, et al., (2018) Paternal grandfather's access to food predicts all-cauyse and cancer mortality in grandsons, *Nature Communications* 9:5124. doi: 10.1038/s41467-018-07617-9.

Verweij M, (1999) Medicalization as a moral problem for preventive medicine, *Bioethics* 13(2):89–113.

von Wright, Georg, Henrik (1997) The Good of Man, in: Carson, Thomas, L., Moser, Paul, K., (eds.) *Morality and the Good Life*, Oxford University Press, New York pp. 147–63.

Yusuf S, Hawken S, Öunpuu S, et al., (2004) Effect of potentially modifiable risk factors associated with myocardial infarction in 52 countries (the INTERHEART study): case–control study, *Lancet* 364:937–52.

Zola, IK, (1972) Medicine as an institution of social control, *Sociological Review* 20:487–504.

Part III

Rethinking professional responsibility

5 On patients' difficulties in understanding medical risks and the aims of clinical risk communication

'They don't really understand'

Ulrik Kihlbom

Introduction

Risk communication and shared decision-making are intrinsic to good medical practice. For example, health care professionals need to clearly explain to their patients the expected benefits of a drug balanced against the risk of adverse reactions. Clarity of communication is especially needed in cases where there are high benefits and high risk related to treatment options associated with treatment decisions that some leukaemia patients face.

Leukaemia is a group of cancers in the myeloid and lymphoid blood cells. In haematological malignancies, the presence of specific somatic mutations has also extensively been used to stratify patients into risk categories that might benefit from risk-adapted treatment, particularly consolidation treatments with autologous or allogeneic haematopoietic stem cell transplantations – i.e. bone marrow transplantation. Clearly, the clinical context of risk communication and consultation with leukaemia patients is very complex. Depending on age, specific mutations and phase in the illness trajectory, many of these patients face complex treatment options that involve different and very severe kinds of risks, especially for leukaemia patients facing a bone marrow transplantation. These decisions should consider both short-term and long-term adverse effects as well as the patient's own attitudes towards risks and counselling. These effects, including life-threatening infections, severe chronic graft-versus-host diseases, and late malignancies, require successfully managing the challenge of communicating risk to already vulnerable patients.

There is an increasing demand put on health care professionals to communicate risks properly in the context of clinical decision-making. Communicating risks accurately has been the subject of a recent change to British law following the *Montgomery v Lanarkshire* [2015] ruling by the UK Supreme Court. Before this ruling, valid consent was satisfied if a reasonable body of medical opinion would agree with the risks presented to the patients by their health care professionals. British law now states that physicians are required to take 'reasonable care to ensure that the patient is aware of any material risks involved in any recommended treatment, and of any reasonable alternative or variant treatments'. This requirement means that clinicians should sufficiently communicate risk so their patients can make a decision that reflects their own values; therefore, health care professionals are required to understand what their patients consider to be a material risk. This same trend can be seen in many other countries (for example, in Sweden and the USA).

What does it mean for a patient to sufficiently understand a medical risk when making a treatment decision associated with high risk and high benefits in clinical practice? That is, what level of understanding is ethically required for making an informed decision when considering risk in a clinical context? According to Beauchamp and Childress' influential account of patient autonomy and informed consent, sufficient understanding of informed consent means that 'one has justified beliefs about the nature and consequences of one's actions'. They claim that full understanding is not required, but they point out that a single missing fact can deprive a person of sufficient understanding (Beauchamp and Childress, 1989, p. 69). Although plausible, this definition is not very helpful in clinical practice.

The starting point of my discussion will not be a theoretical account, such as the one suggested by Beauchamp and Childress. I start by examining how patients understand risk as expressed by haematologists. Haematologists are concerned that newly diagnosed patients may have a difficult time really understanding the severe side-effects of certain treatments. My starting point is material collected in a pilot interview study with eight haematologists at Uppsala University Hospital and Karolinska University Hospital, Sweden, performed by the present author in spring 2016. The interview guide consisted of questions that addressed three themes: risk information, directiveness and relatives. The participants related that leukaemia patients had a difficult time understanding the possible side-effects of bone marrow transplantation. As can be seen in Table 5.1, the haematologists also thought that in most cases they should be direct and use qualitative statements such as 'low risk' and 'very high risk'. This direct approach, however, is not shared by genetics counselling where it is often suggested to be non-direct with patients (Wolff and Jung, 1995). Moreover, the use of qualitative rather than quantitative risk estimates may be in conflict with the evidence of how to best communicate risk. I will return to both of these themes at the end of the chapter as they seem to relate to the first theme in a significant way.

The purpose of this chapter is not to uncover what the haematologists actually meant in the interviews. Rather, the aim here is to discuss and elaborate on ethical norms and concerns that may be in play in clinical

Table 5.1 Themes and sub-themes of the pilot study

Themes	*Sub-themes*
Understanding of severe side-effects in risk communication and decision-making	Patients' difficulties of understanding The importance of understanding Pre-empt criticism
Recommendations	Little hesitation about giving recommendations Sometimes patients must decide 'Doctor, what would you do if you were in my shoes?'
Risk is preferably communicated with qualitative statements	Risk estimates concerns groups, not individuals Numbers diverge attention

practice in the context of risk communication and medical decision-making. More precisely, the following three questions will be addressed: (i) What do haematologists want their patients to understand?, (ii) Why would this understanding be important?, and (iii) Is this concern justified?

This chapter is organized as follows. The following section elaborates on the first of two interpretations of how patients understand risk according to their haematologists. The first interpretation is called the phenomenological interpretation: an intuitively attractive approach that suggests a strong understanding of risk. The second interpretation, outlined in Part III, is weaker and different kind of interpretation – the appropriate attitude interpretation. This interpretation can more easily be related to the motivations that some of the haematologists seem to have for their worry. In the fourth section, the motivations that seem to underpin their worry are outlined in terms of the normative notions: *autonomy, responsibility and regret*. Lastly, a consideration is briefly elaborated that may explain why it is so hard to fully understand the risks that patients may face.

The phenomenological interpretation on the required understanding of risk

The pilot study revealed a main theme: leukaemia patients had difficulties understanding the side-effects of treatment. For example, one interviewed haematologist expressed this theme as follows:

> I think it is terribly difficult to understand how you feel after a chemotherapy treatment if you haven't experienced it before [...] the information bit and discussion are much easier with an experienced patient. [...] [T]here are some diagnoses that go directly to transplantation without previous treatment and then try to [...]. I think I've

> seen that through the years that those [patients] have much more difficulties in understanding what it is about.
>
> (H5)

As the above quotation expresses, it was typically the side-effects of the treatment and the difficulty of understanding them rather than the risk of death that the haematologists were concerned about. Among the severe side-effects and conditions that the haematologist is referring to are long-term fatigue, serious infections, nausea and vomiting, stomach cramps, diarrhoea, pain, and weight loss. These can be extremely severe and long-lasting effects associated with suffering and low quality of life. This issue is expressed in the following quotation:

> The problem is often not that [the patients] do not understand that they may die as a result of the treatment, that is rather [...] we tell them that 'you can die from this treatment or be cured'. And I believe most of them rather easy can [...] maybe not emotionally understand, but they can know that they have heard it, that it has entered their heads so to speak. But the morbidity that you can get so incredibly ill goes beyond everybody's understanding because you cannot understand it if you haven't seen it. Even if you try to explain, it surpasses their world of imagination.
>
> (H4)

Here, the informant expresses a stronger view on the difficulty if not impossibility of understanding the seriousness of the side-effects. This difficulty of understanding was emphasized or brought up by all interviewed haematologists. The weight that they put on this difficulty relative to other challenges in the risk communication was noteworthy. They were less concerned with understanding probabilities or weighing the benefits and harms in relation to the probabilities. In the scientific literature, these are topics that receive massive attention. Moreover, even if several of them highlighted the dread-factor of cancer (i.e. that receiving a cancer diagnosis may block normal practical reasoning), the difficulty was in emphasizing what it would be to like to experience the side-effects.

Why do the experts think it is so difficult to 'really understand' the side-effects? Most of us have experienced bad stomachs, irritated skin and nausea, at least to some extent. Would it not be possible to just imagine more intensity and suffering related to these conditions? One possibility is that the understanding the haematologists have in mind is an understanding from the inside (i.e. from the privileged position of the subject experiencing the event). For example, often people make statements like 'unless you have experienced it, you cannot really understand what it is like to give birth', 'what it is to live with autism', or 'what it is like to be in the middle of a real storm at sea'. It is certainly not the intention to ridicule such statements here. On the contrary, they often seem to contain some truth, so this phenomenological view is intuitively attractive. Typically, direct experience of a very intense or traumatic event has a significant emotional impact. As people's imaginative powers are limited, it is difficult (if not impossible) to provide a perfect simulation of the input our senses provide.

Some of the haematologists seem to ideally want their patients to grasp this kind of understanding, although realize that it is probably out of reach. This interpretation will here be called the phenomenological view. Phenomenology can, very briefly and generally, be said to be the philosophical study of conscious experiences experienced from the first-person point of view. A phenomenological understanding of really understanding side-effects is then here to understand what it is like for that person to experience those side-effects. If the experience in question is of an unusual kind and totally new, it might obviously be difficult to reach such an understanding. One might even argue that if having a certain understanding is identical with having a certain experience, then only persons actually having the experience can have the understanding. This may imply that even the people who at one point had an understanding of this kind will probably not have the same understanding at a later point since memories are deceptive.

If this is the kind of understanding that would be required to consent to medical treatment (e.g. in the case of leukaemia patients to bone marrow transplantation), it is very difficult, if not impossible, to consent to something you are not experiencing. This view might be correct in that there might be an unbridgeable gap between the ideal understanding and what risk communication in the clinical setting can convey, or so it seems to me.

The appropriate-response interpretation – understanding as evaluation

One of the haematologists said during the interview:

> [Patients opting for a bone marrow transplantation] shall know what they venture into. And think that the risk–benefit relationship is acceptable.
>
> (H4)

The second sentence in this quote suggests that newly diagnosed leukaemia patients facing bone marrow transplantation should not only given information that allows them to understand that they face very serious side-effects and what those are, but also to evaluate and accept them.

Although it may be true that it is impossible to imagine what it would be like to have a certain side-effect by simulating the experience of the particular side-effect, it might possible to understand what it is to have the side-effect given that understanding is to have a grasp of the meaning of having such an experience (rather than to have the experience). In such a case, we might find a plausible interpretation that does some justice to the intuition that without the experience it may be quite difficult, but in principle possible, to 'really' understand what it would be like, for example, to have severe skin rashes for several years.

What kind of meaning is in use here? A beginning of an answer is to draw a distinction between cognitive meaning and evaluative meaning that allows us to talk about grasping cognitive meaning versus grasping evaluative meaning. There are a vast number of quite different theories of meaning, but one influential view says that grasping the *cognitive* meaning of a statement sometimes is to grasp the truth conditions of the sentence (Davidson, 1967). To illustrate, if someone says: 'It is raining', then you understand that utterance if you understand under which condition that utterance is true. Grasping the *evaluative* meaning involves arguably more than that. According to one view on evaluative meaning, understanding 'for instance' statements about medical risks is rather to have *an appropriate attitudinal response* to the event(s). (McDowell, 1998; Wiggins, 1987; Kihlbom, 2018). On this appropriate response-view, evaluations should be understood as attitudinal responses to actions, events and situations etc. Responses as evaluations may be judged as normatively appropriate or inappropriate. Suppose a haematologist, while describing the side-effects associated with bone marrow transplantation, says that the risk for serious side-effects is very high. On the appropriate response-view, the patient understands that statement if they respond to the statements in an appropriate or fitting way. However, the view is silent on what an appropriate response amounts to in a specific situation. In our everyday thinking about attitudes toward risks, we do think, correctly in my view, that people may underestimate or overestimate risk or may make poor judgements about what is good or bad. For example, information concerning considerably increased risk for severe side-effects of a certain treatment may cause a patient to decline the treatment. Patients who fail to respond in a similar vein may do so because they fail to conceive what is at stake (i.e. they do not cognitively or evaluatively understand their situation). This view on evaluations and values is a form of realism – i.e. there are values 'out there' that we may apprehend. However, values can only be understood with reference to how a person should respond to them, something that may be a contested issue. In this view, there is no other way to determine what is appropriate than to argue for one's view. It is, in this sense, a weak form of realism (McDowell, 1998; Kihlbom, 2002).

If we employ the appropriate response-account of evaluative meaning when interpreting the haematologists, then to 'really understand' is rather to appropriately evaluate one's situation – i.e. properly respond to the side-effects of the bone marrow transplantation. In other words, this leads to the interpretation that newly diagnosed leukaemia patients facing bone marrow transplantation have a very hard time properly evaluating and responding to their situation. That a response is merited implies that there are reasons to respond in a specific way. So, the notion of evaluation as a merited response introduces an ethical normativity with regards to patients' understanding side-effects. However, it is important to note that the response of the patient is not the only normatively relevant aspect here. From this view, there are mainly two sources of normativity regarding the context of a clinical consultation of a bone marrow transplantation. On the one hand, there is the haematologist's description of the possible scenarios; on the other hand, there is the patient's response to the description. Both the haematologist's description and the patient's response are inherently normative. Descriptions of potential risks are not normatively innocent, as they may function as reason to decide or act in a certain way. To 'really understand', one must reasonably have a correct and sufficiently rich description. In the clinical setting, viewing evaluations of side-effects in this way makes room for negotiations in terms of discussions of how to describe the options and how to respond to the options.

While these may be the most plausible interpretations of what some of the haematologists meant by 'to really understand', it is compatible with the phenomenological interpretation. That is, clinicians might hold the view that an ideal understanding of potential side-effects of bone marrow transplantation is one along the lines of the phenomenological interpretation, as well as aiming at having the patient evaluating the side-effects in an appropriate way. Why is it then that the haematologists worried that leukaemia patients do not understand the side-effects in one or both these ways? In the following section, I will outline several possible motivations.

Patient autonomy, responsibility and regret

As said previously, the appropriate response interpretation introduces ethical normativity into the context of risk communication. One strong ethical norm in the clinical setting is patient autonomy. Here are some quotes that point in that direction:

> [T]his is about the patient and he or she must have the final say. The patient cannot demand an unreasonable treatment [...]. But the patient can say 'no' to a risky and potentially harmful treatment and choose a less risky alternative, and the patient needs information to make such a choice.
>
> (H3)

> [Transplantation] might be the only chance for a cure, but it might still not be the best option [...] and [I] try to present all treatment options as viable alternatives, because I think you have a right to know.
>
> (H7)

They seem to be saying that patients are entitled to be informed about the risks in order to be able to make decisions based on sufficient understanding and in agreement with their own values – to be able to exercise their autonomy.

This approach might help the haematologists align their concern that their patients have difficulty understanding the side-effects of bone marrow transplantation with what legal requirements and guidelines require *vis-a-vis* patient autonomy. Indeed, patients' lack of understanding seems to part of their worry, but it might not be the only one. There are a few other aspects that are both interesting in themselves and relate to the difficulty of understanding the impact of side-effects. For example, if patients autonomously make decisions about risky treatments (i.e. on the basis of being sufficiently informed and having the capacity to imagine and evaluate the different options), they can also be held responsible and accountable for their choices. To be held responsible means that one might be unfavourably judged if these requirements are ignored or flouted (Watson, 1996). To promote patients' autonomy in risk communication may also mean that patients can be held responsible for consenting to treatment. Why would it be important to the haematologists that patients can be held responsible or accountable for their consent and who would actively criticize patients' consent? In Sweden, clinicians are legally responsible for medical decisions and for offering treatments. None of interviewees suggested a view that health care professionals would or should hold the patients responsible. However, one can also consider the possibility that the haematologists aimed to have their patients hold themselves responsible: 'And for compliance [...] the patient needs to be informed and motivated in order to endure severe side-effects [of treatment]' (H3).

In our pilot study, as noted above, the haematologists thought that they often should recommend a treatment option. One notion of autonomy that seems apt here is Rebecca Kukla's notion of conscientious autonomy:

> When we act conscientiously, we act out of a commitment to diligently uphold principles, practices, goals, values, or regimes that we take as our own normative standards. Conscientious action stems from a responsible commitment to the norms that govern it.
>
> (Kukla, 2005, p. 38)

So, a choice by a patient reflecting autonomy might be best understood as a commitment to certain norms, partly those regulating medical practice. As Kukla points out, it might also be better understood as ongoing reflective practice rather than a momentous or discrete decision. This notion of patient autonomy acknowledges the normative commitment that patients make towards certain practices even if such practices are recommended by others, including people in an authoritative role such as physicians.

Therefore, leukaemia patients may autonomously accept the haematologists' recommendations and directives for further treatment.

A less ethically flattering possible motive for the clinicians would be that they want to avoid criticism from patients and their relatives in situations where the patients suffered difficult side-effects. Here are a couple of quotes that support such an interpretation:

> [I]t can become very difficult if they get problems after [the transplantation], if they think that they were not informed about the risks.
>
> (H6)

> [I]t is like [...] you know, I have seen patients who [said] 'I wouldn't have consented to treatment, had I known this.' They have had such long-term complication that lower their quality of life.
>
> (H3)

Here, the haematologists seem to be expressing the desire to avoid patients criticizing or regretting the choice of treatment. Indeed, it is fully understandable that clinicians, as most people, want to avoid criticism. Although avoiding criticism is not the motive that should drive risk communication in the clinical setting, nothing suggested that this was the motive. Indeed, the avoidance of criticism interpretation might be too simplified, at least as the only explanation of the haematologists' concerns. The above two quotations indicate that a key motive is to avoid regretting decisions on behalf of the patients. Several of the haematologists said that they were concerned with possible future regret of the patients and their well-being if things turned out badly. This view is in line with previous research – i.e. health care professionals are influenced by their concern that patients may suffer from regret, for example, after decisions about sterilization and sex change (Mertes, 2017; McQueen, 2017).

Understanding and referential opacity

The remaining question for this chapter is if the worry of the haematologists seems justified. There are many reasons for why it might be very difficult for a newly diagnosed leukaemia patient to 'really understand' the possible side-effects of a bone marrow transplantation. The notion of risk, the dread factor and other psychological effects are some known barriers to successful risk communication. However, the problem I would like to address here is a more fundamental problem that is really hard to overcome and that is underestimated in the literature – the nature of consent. A possible cause for the latter is that this philosophical problem is difficult to get a grip of. As Onara O'Neill, and others (O'Neill, 2001) have pointed out, consent is a propositional attitude and as such it is referentially opaque. Our wants, desires or hopes are consent attitudes directed towards particular and a very limited number of descriptions of the world. However, there are an indefinite number of true descriptions, or propositions, of a certain series of events, such as of a medical treatment that results in certain side-effects. Suppose that I hope for a new job position and to my joy I get it. But, alas, after some time it turns out that the job was not what I hoped for. As a matter of fact, I cannot stand the new job. My negative experience might be so even if my previous understanding of the job was correct. It was just that other descriptions or propositions of the job were also true, but these descriptions were unclear or opaque to me before I actually engaged in the work. This referential opacity holds for consent to medical treatment, too. That is, a description of a treatment is not necessarily or automatically transferable to another or a more specific description of the treatment in question

This may seem an abstract philosophical problem, but its impact for practices such as informed consent in health care is, as O'Neill says, massive (O'Neill, 2001, p. 692). That is, referential opacity may explain why we see reports from clinicians such as from H3 above or from this physician in Australia:

> Sometimes afterwards patients will say, 'If I'd known it was like this, I wouldn't have done it.' I say, 'Well, we talked about it, I showed you the pictures, I said you can die from this, you can't get much sicker than that' and they say 'Yeah, but I just didn't realize it.'
>
> (Jordens and Montgomery, 2018, p. 20)

The problem of the referential opacity of informed consent cannot be counterbalanced with less frugal or more specific information. It is a genuine problem relating to the principled limitations of our language to describe the world. As such, it may be one explanation of why the gap between the ideal understanding of possible side-effects of bone marrow transplantation and communication in the end seems unbridgeable – our language and mind can but only reach a fraction of possible future experiences.

Conclusion

To communicate serious medical risks to patients in clinical consultation is challenging for a number of reasons. The outcomes might be daunting as well as uncertain, the information is highly complex, and then there is the more fundamental problem of referential opacity. But clinicians have also further, fully legitimate, aims around communicating risk than to provide factual information of medical prognosis and risks. If the analyses above are on the right track, then it seems that the normative aspects of risk communication are crucial in such risk communication. That is, clinicians might primarily be concerned with preparing the patients for what lies ahead, how uncertain, ambiguous and grim that future may be. Patients' normative attitudes with regards to the possible impact of the side-effects, personal autonomy and future regret are important elements of the understanding. Moreover, the reasons why the haematologists in the pilot study were reluctant to use specific descriptions and risk estimates in their communication with leukaemia patients may very well relate to this aim of preparation. Specific descriptions and/or precise risk numbers might have an undesired effect, namely that patients might stick to that outcome rather than to have a more open mind towards the vast number of possible consequences. It seems to me that these normative aspects that are somewhat overlooked in the bioethical discussion and in need of further attention.

References

Beauchamp T., Childress J., (1989) *Principles of Biomedical Ethics*, 3rd edn. Oxford: Oxford University Press.

Davidson, D., (1967) Truth and meaning. *Synthese*, 17: 304–23. https://doi.org/10.1007/BF00485035.

Kihlbom, U. (2002) *Ethical Particularism*. Stockholm: Almqvist and Wiksell.

Kihlbom, U. (2018) 'Genetic Risk and Value', *Journal of Risk Research*, 21(2): 222–35.

Kukla, R. (2005) 'Conscientious Autonomy: Displacing Decisions in Healthcare', *Hastings Center Report*, 35(2): 34–44.

McDowell, J. (1998) *Mind, Value, and Reality*. Cambridge, MA: Harvard University Press.

McQueen, P. (2017) 'The Role of Regret in Medical Decision-making', *Ethic Theory Moral Prac*, 20: 1051–65. https://doi.org/10.1007/s10677-017-9844-8.

Mertes, H. (2017) 'The role of anticipated decision regret and the patient's best interest in sterilisation and medically assisted reproduction', *J Med Ethics*, 43: 314–18.

Montgomery v Lanarkshire Health Board [2015] SC 11 [2015] 1 AC 1430.

O'Neill, O. (2001) 'Informed Consent and Genetic Information', *Stud. Hist. Phil. Biol. & Biomed. Sci.*, 32(4): 689–704.

Jordens, C. and Montgomery, K. (2018) 'Respecting Patient Autonomy Some Telling Challenges for Medical Professionals Who Treat Seriously Ill Patients', in *Medical Professionals Conflicts and Quandaries in Medical Practice*, Montgomery and Lipworth (eds.), New York: Routledge, doi: https://doi.org/10.4324/9780203712221.

Watson, G. (1996) 'Two Faces of Responsibility', *Philosophical Topics*, 24(2): 227–48. doi: 10.5840/philtopics199624222.

Wiggins, D. (1987) 'A Sensible Subjectivism?', in *Needs, Values, Truth: Essays in the Philosophy of Value*, Wiggins, D. (ed.), 185–214. Oxford: Blackwell.

Wolff, G., Jung, C. (1995) 'Nondirectiveness and genetic counseling', *J Gen Couns*, 4: 3–25.

6 Ethical consideration about health risk communication and professional responsibility

Silke Schicktanz

Introduction

It can be said that *where there is risk there is responsibility*. The work of the philosopher Hans Jonas on 'the imperative of responsibility' (1984) has laid the foundation for a theoretical union of risk and responsibility in applied ethics. Jonas referred to different types of technical risk, emerging from nuclear power or the genetic engineering of humans. According to him, traditional ethics oversees the particular challenge of risks implied by those technologies. Both the action and the event have individual *and* collective dimensions because the event is probabilistic and its time dimension is often beyond the imagination of an individual's lifetime. According to Jonas, the main ethical problem associated with modern technology in the life sciences is the manipulation of and risk to the *conditio humana*. In contrast, using modern technology to predict future risk is understood to be a helpful or even necessary condition required for acting responsibly about the future of the individual and the collective. However, it was not so obvious at the time that new ethical problems might also arise from modern predictive technologies such as genetic testing. The ethical dimensions arising from risk prediction and prognosis are manifold. Do we want to know our future health risks, or do we even want to know our medical fate? What if the opportunities to change the course of disease are limited? With whom shall we share this information? And what if this information is not deterministic, but probabilistic?

For a long time, these questions have been raised for a few, mainly hereditary conditions, such as Huntington disease, *BRCA* 1 and 2 genes for breast cancer, or the prenatal diagnosis of Trisomy 21. Most common diseases, however, are poly-genetic or are multi-causal, as genetic and environmental factors must come together. Only in the last decade, have the technical opportunities for producing this complex type of risk information grown substantially and quantitatively because of the invention of advanced biostatistics and improvements in data management. The new paradigms of Big Data and machine learning promise a radical new approach to deal with all kinds of health risks. In the near future, so is the promise, we will have so much more information about the individual risks for complex diseases, based on the combination of genetic and non-genetic data. Health risk information combines factual (statistical and probabilistic) information (how likely is a disease/phenotype) *and* value information (which event/phenotype is considered a negative event for human health?) (see also Hansson 2010). For some this risk information is seen as helpful to plan later life or to consider a healthier lifestyle. For others, this information is a social and emotional burden. It can lead to stigmatization and is discrediting.

The options of knowing one's own complex health risk are still limited, but a few examples, such as the current debate of predicting Alzheimer's disease, provide a timely example for the related ethical and social dimensions. Of course, the extensive debate about genetic responsibility also provides a helpful starting point to examine the normative dimensions of responsibility (Leefmann et al. 2017; Schicktanz 2018).

In this chapter, I will explain first what I mean by 'responsibility' and how this understanding provides us with a very detailed view on the various normative spheres relevant for health risk communication. In the analysis part, I will employ three exemplary cases of health risk information and show how they enfold different dimensions of responsibility. Hereby, I focus on professional responsibility in communicating the different types of risks. The different examples illustrate that the professional responsibility of communicating risks entails not only factual and numeric competencies, but should also require systematic reflection on the norms and consequences underlying the assumptions of each risk communication.

A proposition to understand responsibility in applied ethical contexts

The concept of responsibility has undergone its own bioethical evolution (see Schicktanz and Schweda 2012 for further detail). It is important to note that there appear to be three phases in the evolution of the notion of responsibility in the field of bioethics. The first, from the 1960s onwards, was dominated by discourses on *collective forward-oriented responsibility*, often directed towards the next generation, humankind in general or nature. Prominent advocates of the ideas from this phase included Hans Jonas and also Van Rensselaer Potter (1970) (see also Whitehouse 2003); both can be seen as founders of bioethics as a discipline. The second phase, starting in the mid-1970s, began as an intensified discussion of *professional responsibility towards individuals*. The American *Belmont Report* of 1976, which is seen as a touchstone for informed consent procedures and the legal protection of human research participants, precisely defined the responsibilities of individual researchers. Professional responsibility was understood as both a past-oriented model of liability ('who did morally wrong?') and a future-oriented model of guidance ('what shall we do?') by stressing the important, powerful role ascribed to professionals. The third phase, starting in the 1990s, deals with the *interrelationships between social and individual responsibility*. This continuing trend can be seen as a reaction to political reforms in which public welfare and health care systems were successively cut back (Ter Meulen and Jotterand 2008; Young 2011). However, the controversial character of knowledge and of risk assessment added a new dilemma to the discussion of social and individual responsibilities. While health and risk information is seen as empowering and as a means of reducing illness and social costs, it also burdens the individual as they can now be blamed for becoming ill (Yoder 2002).

As well as its philosophical-ethical usage, the notion of responsibility has also become central to the approach of the social sciences to genetic risk since the turn of the millennium. In sociological studies of genetic risk, responsibility is now termed 'genetic responsibility', which has become a key term used in the description of moral imperatives (Leefmann et al. 2017). The new trend to focus not only on genetic risks, but on all kind of health risks, especially those that might impact age-related or chronic diseases, allows for expanded meaning, namely 'health responsibility'. Health responsibility covers a broad spectrum of ethically relevant issues: from the provision of (effective) primary prevention to the risk that citizens will be accused of being 'guilty' of having lived an unhealthy lifestyle 'causing' the disease. When discussing genetic risks, various authors have pointed out the absurdity of claiming responsibility based on the fact, that genetic risks are involuntary (Lemke 2006). However, genetic risk information can impact behaviour, treatment decisions or family planning (Hallowell 1999; Featherstone et al. 2006). Therefore, a 'new' genetic responsibility regarding the disclosure or non-disclosure of genetic risk within families has been observed.

The historical and sociological usage of the term 'responsibility' reveals a deep discrepancy. On the one hand, the word is highly relevant in public and academic debates but, on the other hand, its meaning and implications are quite often far from being self-evident. To bring some clarity into this debate, we have developed and further developed an analytical tool for understanding responsibility (Schicktanz and Schweda 2012; Schicktanz 2018). This tool can be employed very practically to critically reflect the existing practical and theoretical approaches to responsibility in mundane discourse (e.g. on genetic responsibility as by Schicktanz 2018 or Wöhlke and Perry 2019; or as by Schweda and Pfaller 2020 on ageing medicine).

The dominant use of the term 'responsibility' in the legal sphere has shaped a reductionist usage of responsibility in a manner suggesting 'being blameworthy' or 'being guilty' (Holl et al. 2001; Hart 1968). But moral philosophers have attempted to develop a typology for responsibility going beyond this narrow understanding (e.g. Baier 1991; French 1991; Bayertz 1995). This research has revealed that the formulation of 'responsibility' comprises a set of relationships between different positions (relata). Such a meta-ethical understanding promotes the idea that the conditions and normative assumptions of each relatum, as well as their relationships with each other, can be analysed. It constitutes a *general conception* of how to compose or analyse a morally plausible statement about the relationships between several entities. The most basic concept of responsibility requires at least three relata: someone (the *moral agent*) is responsible for someone or something (*the moral object*) against someone (*moral authority*). On closer inspection, it turns out that even more – five, six or even seven – relata may be necessary in order to reconstruct and analyse usage of the concept 'responsibility' in an adequate manner. For the detailed analysis of the role of responsibility in the different fields of bioethical debate, we suggest a concept involving seven relata; i.e. someone (the *moral agent*) in a particular time frame (the *time*) is retrospectively/prospectively (the *temporal direction*) responsible for something/someone (the moral *object*) that is overseen by someone (norm-supervising *authority/enforcement*) on the basis of certain standards (*norms*) with certain consequences (the *sanctions or rewards*) (see Schicktanz and Schweda 2012).

There are at least four basic axes running throughout this formula, which can provide valuable perspectives for further ethical analysis: the agent–object relationship, justification of norms and consequences, the epistemic, and the time dimension. These four axes provide the analytical tool to examine in detail responsibility ascriptions.

The agent–object relationship: the moral agent–object relationship is historically, culturally or socially embedded and constrained by social norms, role expectations, power divisions, emotions (e.g. trust or fear) and knowledge biases. A good example of how this relationship can frame our moral assumptions is reflected in the moral dimension of the doctor–patient relationship, or the parent–child relationship. The agent (doctor/parent) is in a particular (social, perhaps also cognitive and emotional) power position, while the object (patient/child) is dependent on but also trusts the other. Responsibility is associated with the competence and/or the power to decide (Bayertz 1995). One important argument that supports the idea of collectives as moral agents, too, points to our intuition that *sometimes* the whole (a family, a corporation) is more than just the sum of its single parts. Those social groups, such as the family but also the state, a patient association or a health care insurance company, can be seen as moral agents if their actions are based upon a collective mind or an '*esprit du corps*' (Beier et al. 2016). This has to be understood as a joint commitment, and its effect cannot simply be explained by the aggregation of individual actions (French 1991).

Justification of the norm–authority–consequences relationship: the category *norm* refers to normative standards such as moral principles, and social or legal norms. Practical norms do not stand alone – at least in a coherent, consistent justification but are related to theoretical presuppositions of an authorization. For example, the moral norm 'you must not kill' is related to assumptions whether this norm is universally valid (due to rational, secular standards such as human rights, but might also include exceptions) and connected to social or legal consequences (in legal terms, this can be a court sentence in a case where there is evidence that person X killed person Y, in social terms it can be a social group or 'society' who express their moral disgust). The court or law function as a norm-supervising authority with clear consequences (a sentence or a fine), while in the case of social norms, the ambiguity of social structures as normative authority can result in very different consequences (e.g. social exclusion, social conflict or ignorance). Individual moral norms function as commitments for individuals and, hence, our conscience is the authority. The consequences vary according to the practical power held by the 'authority': from feelings of being morally good or guilty, to social or legal consequences. Hence, the consequences can include not only sanctions for unwanted events but also rewards for good outcomes. It would be expected that the more binding and important an underlying moral norm is, the more demanding or powerful is the sanction or reward for the respective event.

The epistemic or causality dimension: how much is an agent free to determine the outcome of his/her action? Despite the unsolved theoretical battle between compatibilists and incompatibilists, our moral practice to ascribe responsibility refers to clear causal ascriptions of whether the behaviour of a person has led to a particular outcome of the situation. In this sense, where there are doubts regarding the causal links, persons can be excused or exempted from blame. With regards to health, a crucial question is how far behaviour determines a certain health condition. The scientific basis for this assumption is often rather vague.

The time dimension: in the classical legal context (Bayertz 1995), responsibility was primarily discussed *retrospectively (backwards)*. Guilt as responsibility is then used to hold a person accountable for harmful events that happened in the past. This has been referred to as the liability model (Young 2011). However, as modern society developed, a more *prospective* meaning became prominent (Bayertz 1995). According to this idea, having responsibility for an object, a person, or a group means 'being in charge of' them or 'taking care of future events'. The time dimension also points to the often difficult but important question of whether an event is a short- or long-term event (next week or in 20 years?) and how the time scale is assessed according to our life course (Schicktanz and Schweda 2012).

The way I use responsibility here differs from the broad sense as proposed by Strawson (2008). Instead, I propose to focus its usage on moral statements that refer to concrete relationships between a moral agent and a moral object in an established, socially recognized relationship (e.g. a partnership, parenthood, citizenship, the doctor–patient relationship). All these relationships are very conditional for the analysis as they constitute the epistemic, cultural, power-related and emotional underlying premises which need to be made more explicit.

The emergence of health risk communication in modern medicine

In the following, I will illustrate how responsible health risk communication entails several ethical considerations by using three prime examples of health risk communication. My analysis focuses on the question of professional responsibility, namely the responsibility of those in power when it comes to the provision of health risk information. Hereby the term professionals refers not only to the individual doctor, as in doctor–patient-communication. I also consider public health experts or scientists as members of the group of professionals that are responsible for health risk communication. It might depend on the respective case, which group exactly is in charge, but often it is a web of professionals who create and shape information and communication.

My first example refers to public health communication regarding *breast cancer risk* and how it is communicated to women. The persons in charge that I have in mind, are experts in publicly funded health institutions, such as ministries or administration. Public health communication is multi-professional, as here not only doctors are involved, but also persons in public health administration, media experts, or sometimes private companies that have been awarded a contract by the public institution. The second example refers to the new paradigm of 'personalized' or *stratified cancer treatment* by employing so-called genetic biomarkers to tailor cancer treatment. The third example refers to the emerging debate around '*new dementia*'; where the combination of newly available risk information regarding late-onset Alzheimer's disease triggers claims for taking preventive measures on the individual side.

Public health communication about knowing women's breast cancer risk

Breast cancer is one of the most common cancers. Many countries developed public health strategies to motivate women to attend regular check-ups and to increase awareness about 'breast cancer' as a public health threat (the so-called pink ribbon movement). Various studies have shown that these measures have impacted the perceived risk in women: The perceived lifetime risk of developing breast cancer in studies varies between 30 and 46 per cent, much higher than the real lifetime risk of 12 per cent (de Jonge et al. 2009).

This includes publicly financed screening programmes for detecting pre- or early stages of breast cancer, or more general, health communication campaigns to promote a healthy lifestyle to prevent breast cancer. The 'healthy' lifestyle tends to repeat the 'usual suspects' such as: stopping smoking, drinking less alcohol, eating more fruits and vegetables, reducing weight.

Our current knowledge about rare genetic variants that increase the risk for breast cancer such as *BRCA* 1 and 2 (two tumour suppressor genes) has added to the ethical challenges of adequate risk communication. According to the lay-relevant website of the US NIH, '12% of women in the general population will develop breast cancer sometime during their lives... By contrast, a recent large study estimated that about 72% of women who inherit a harmful *BRCA1* mutation and about 69% of women who inherit a harmful *BRCA2* mutation will develop breast cancer by the age of 80'.[1] Risk-reduction for those who have inherited the respective *BRCA* 1 or 2 mutations, includes several options, such as regular screening, chemotherapy and 'preventive' mastectomy. NIH recommends that counselling and genetic testing should only be done in particular cases with family history because both mutations are rare (2–3 in 1000 women of the general population, and only 10–15 per cent of breast cancer are related to these *BRCA* mutations), and genetic testing should only be done after proper counselling and in case of particular family history (Nelson et al. 2013).

However, the so called 'Angelina-Jolie-effect' has pointed out the particular challenge of public health communication: there is not only professional, expert-guided communication, but also public health communication happens in a specific socio-cultural and media context. After the famous actress Angelina Jolie publicly announced her bilateral mastectomy after it was detected that she is a *BRCA1* mutation carrier this evidently led to an increase in many US women seeking genetic testing and risk-reducing bilateral mastectomy (Liede et al. 2018). Risk psychologists such as Gigerenzer (2014) stressed that public health communication on breast cancer has suffered from inadequate risk communication for a long time. To motivate women for mammography and regular check-ups, pamphlets and websites stated that mammography would reduce the risk for breast cancer up to 20 per cent and even more.[2] The correct risk to be reduced is much less, namely that only four instead of five women out of 1000 will get cancer. The absolute risk is much more informative than relative risk information as Gigerenzer and colleagues pointed out. They also argued that women show high overestimations of the benefits of mammography because of this permanent misinformation (Elmore and Gigerenzer 2005). They criticized leading scientific journals for still allowing such misleading risk information presentation in the abstracts. Overall, the communication strategy Gigerenzer and colleagues suggest is to use only absolute risks, to have numbers and correct data by hand and to withdraw from any value-loaded information such as stating that 'screening saves life' if there were no studies demonstrating a decrease in mortality by the practice. The latter has been revised by a current systematic review which found decreased mortality rates in correlation to organized screening programmes (Zielonke et al. 2020).

Reflecting on the professional responsibility in this example means to address a complex case. The agent–object relationship includes the single doctor (depending on local health care organizations, e.g. a gynaecologist or genetic counsellor) and a woman (not yet ill but going regularly to check-ups). However, one must consider also the public health policy makers who provide doctors with background information and decides to provide such tests and check-ups on the basis of public health service (or not). The woman in such a situation, still not diagnosed with breast cancer, is mostly healthy, and as such is not in a weakened position as a patient who has already received the diagnosis of a serious condition. Still, the woman sees the individual doctor as her direct counterpart and expects him/her to provide the best information. Individual doctors might hand out, rarely detailed, information leaflets about breast cancer. Instead, generally, public websites, health insurance and media reports[3] often serve as communication channel for such health risks.

Professional responsibility regarding informing a patient about the risks of a disease or the risks related to a diagnosis or treatment is captured in the ethical-legal concept of 'informed consent' (Berg et al. 2001). Professional responsibility towards an individual patient is seen as balancing the different norms of respecting self-determination, avoiding harm and ensuring beneficence (Beauchamp and Childress 2008). In the context of genetic counselling, there has long been a special professional responsibility to accomplish *non-directive counselling* during treatment, first developed in the context of prenatal care in the 1960s. In the 1990s, another concept conquered the debate: the right 'not to know' (Chadwick et al. 1997). But the right to know or not to know could conflict with the normative right to know/not to know of others, such as a biological family member, if hereditary conditions are at stake (see also Chapter 7 in this volume). If respecting individual autonomy is understood as the basic right to decide *what kind* of information one wants to have, then the 'new' patient's right not to know can be seen as a historical advancement, questioning a simplistic understanding that 'informed decisions' are always better than uninformed decisions (see Kihlbom 2008). Ignoring offers for breast cancer screening is in many countries, however, not easy. The misleading information that overstates the benefits of screening and understates the risk of false-positive diagnosis, and finally, and the overstated risk for genetic mutations causing breast cancer, can undermine the individual's right to determine what kind of information she wants. Hence, the guiding norm for promoting screening and prevention of breast cancer in the current way is – a closer look reveals – rather a public health interest or, partly, about the well-being of the individual, but is not necessarily respecting the autonomy of the woman. It can be questioned whether the repeated provision of misleading information really respects the patient's right to self-determination and 'personally informed' decision-making. Instead, one can criticize that this practice overrules this right because professionals believe it is still better to undergo screening to detect breast cancer earlier than to risk it going undetected, because there are treatments available. Alternatively, one can argue that the underlying norm should be the moral obligation to increase individual (and social) benefit and to avoid suffering (as many consequentialists argue) because it is in the (rational) interest of all human beings. From this, it follows however, a duty to find out whether we can justify our actions upon the given knowledge. Moreover, the question remains tricky as to whether this is a duty to ourselves or a duty to society. A duty to ourselves or, in other words, a deontological understanding of self-care, would still allow for subjective interpretations of what is my personal understanding of a good life, of what harm (e.g. breast cancer and the effects of different types of treatment) means to me and how do I want to deal with this in a way best for me. This liberal version of understanding self-care seems more compatible with the principle of liberal freedom and self-determination.

These ideals of pure 'information' provision and rational choices to be made can be questioned, however. Studies found that women who were confronted with the facts of overstated risks and the risks of false-positive testing during screening, still stressed that they feared 'under-diagnosis' or that women were not seeking regular check-ups (Waller et al. 2013). This might be due to the fact that breast cancer is seen as something 'treatable' in contrast to other health risk information where treatment options are less available (Raz and Schicktanz 2016, 56ff). In such cases, there is an anticipation of a later regret if one did not use the chance to learn about the risk, because then acting on that knowledge would have been a real option (see also Schicktanz 2017). This anticipated regret also refers to the issue of causality: here it is less the question of how screening causally reduces suffering related to breast cancer (because of earlier detection), but how women see a causal relationship between their former action (to go or not to go to screening) and then being confronted many years later with a diagnosis of an (advanced) breast cancer. If screening is at least good enough to increase the chances of detection of breast cancer in the early stages, then this would serve as a cause for feeling later guilty or not.

Overstating information to motivate (or nudge?) women to go to regular screenings because one believes that there exists duty towards society (e.g. for reducing public health costs or to allow for different allocation of resources) explicitly overruns the liberal basic notion. It also collides with many professional ethics codices which explicitly state that individual rights and benefits come first. If professionals believe that public health interest is an adequate justification, they at least must reveal this norm explicitly and must not use the language of patient's autonomy or claim that this serves to support individual choices.

Informing patients about risks and epistemic uncertainties in the paradigm of biomarker testing for oncology

The second example addresses recent research on so-called 'personalized' treatment in oncology. This approach can be better classified as 'stratification', as it aims to categorize patients by genetic, physiological or biochemical markers (so-called 'biomarkers') into responders or non-responders to treatment, with a high or low risk of adverse effects. The future scenario can look like this: a housewife in her mid-fifties, with two adult children, has just been diagnosed with advanced rectum-colon cancer. The surgeon explains her condition to her in detail and describes the common treatment options including Radio-Chemo-Therapy (RCT) before and after surgery. RCT was to be given in the hope that it would shrink the tumour, allowing easier surgical removal. He additionally announces that the clinic will use a so-called biomarker-test for predicting her individual reaction to the RCT before the surgery is carried out, based on the genetics of the tumour. If the test indicates she would not benefit, one would spare her from the possible side-effects that come along with any RCT and surgery would be performed immediately. However, this means that the risks associated with surgery are greater, since the tumour would remain large, the surgery would be more complicated and there would be a higher risk of recurrence after treatment. Explaining to patients the broader scientific background of 'tailored medicine' is seen by many doctors as a real challenge, given the high complexity of scientific backgrounds that biomarker testing, the statistics and the genetic basis of such testing includes (Hessling and Schicktanz 2012). Hence, the usage of metaphors or short explanations is unavoidable in clinical communications. However, even if these are not strategically used, it is very likely that the patient's decision to agree to such a test is framed by words like 'optimizing', 'personalized' and even 'genetics' (Perry et al. 2017). The fact that patients often do not ask for further explanation or express their ignorance of the details can be seen as trust in the doctor (or sometimes it results from fear and dependency towards the authority of doctors, see Wöhlke et al. 2018). Therefore, clinical communications must be seen as a core field of professional responsibility, even if it remains yet vague in many situations as to what exactly this entails or requires.

In this example, the genetic risk information is not about a hereditary condition or a genetic disease. It is about the genetics of a tumour and is based on local tissue mutations. However, it is still a part of the patient's body. The risk involved is therefore not related to the likelihood of the patient having a disease but is instead linked to the likelihood that the tumour will not react to a particular intervention (pharmaceuticals or radiation, for example). Whether the patient is aware of this scientific difference is a challenge that should be addressed. Our interview study with patients revealed that most patients do not understand the difference, but believe that the test is about their genetic situation in general and will therefore be informative for their children as well (Wöhlke et al. 2018; Perry et al. 2017). From this, it can be seen that there is a need for both professionals and the wider society to take responsibility for empowering patients to better understand such details. As such, the biomarker test provides risk information regarding the prediction of the course of the disease under treatment, but not benefiting from a cancer treatment is still a major risk for most cancer patients. This risk information provides to our understanding only an added value for the patient if patients are also informed a) about the risks associated with withdrawing from some special treatments and b) offered earlier in the process sufficient information of alternative ways of palliative care – as biomarker tests indicating no benefits from neo-adjuvant RCT might also indicate lower chances for any efficient cancer treatment.

Prediction and prevention of Alzheimer's disease: professional responsibility to avoid self-responsibilization

The third example refers to current developments in Alzheimer's disease (AD) research where we observe a shift from cure to prediction and prevention, based on a new conceptualization of dementia as a continuum (Schicktanz et al. 2014). This AD continuum theory promotes three stages of a slowly progressing disease with a first, long asymptomatic phase that starts without any symptomatic changes and only pathological molecular deviations become apparent. The disease, secondly, turns into a symptomatic stage involving Mild Cognitive Impairment (MCI) and objective deviant, physiological biomarkers (such as amyloid plaques or brain volume reduction, especially of the temporal lobe). Eventually, in a third stage, it develops into a clinical syndromic disease with an already advanced pathology (Hampel et al. 2014). This 'new dementia' has led to activities in developing various pre-symptomatic testing. The latest news announced blood tests soon to be available to predict AD 10–15 years in advance. PET amyloid scans are also examined for predicting AD but are seen as too expensive and invasive for public screening. In addition to this new trend, a new topic has arrived on the horizon of dementia studies, namely 'prevention', or with a bit less hype: 'risk reduction strategies' (Livingston et al. 2017; World Health Organization 2019; Leibing and Schicktanz 2020).

Previously available genetic testing for familial AD (the 'APOE4' allele test) has led to a rather restrictive, paternalistic approach (Schicktanz and Kogel 2014). It implied a professional recommendation not to test. Based on the assumption that there is no efficient treatment for dementia, professional responsibility has been seen in protecting patients against harms related to psychosocial implications of a predictive dementia diagnosis, such as stress, suicide or discrimination. In this sense, the debate has tried to fix the dilemma of predicting late-onset dementia by focusing on the professional duties of proper information, high-quality counselling and well-reasoned communication that discourage people who want to be tested. While still no efficient treatment for AD exists, with the promise of prevention the whole debate now seems to shift (Leibing and Schicktanz 2020). Knowing one's risk for later developing AD can help, so the hope is, to alter social, health care and lifestyle structures to change the continuous 'course' of the (not yet existing) disease. This raises indeed important questions about causalities and the epistemic quality of statements made. As indicated by Brasure et al. (2018), there exists a considerable uncertainty or ambivalence regarding the promises of dementia prevention. On the other hand, the hope and moral prospects of prevention, often regarded as cheaper, less invasive measures, and promoting well-being are very appealing. So, what kind of professional responsibilities occur when prevention is promoted? Various experts reiterate epistemic vagueness regarding prevention claims by emphasizing that effects found by existing prevention .... Or they point out that study conceptions are not designed appropriately ('too short, too small samples') for making solid claims. However, even if the evidence is partly missing or critically discussed, an underlying moral assumption for many professions seems that recommending a healthier lifestyle is nothing that harms – even if it would be ineffective (Schicktanz 2020). But is that so?

According to Ayo (2012) health promotion and preventive imperatives are especially problematic within a dominantly neoliberal climate. The responsibilization of the individual is a logical result of neoliberal health policy. Minimal state governance and implementation of market rules, as well as a focus on choice-risk management problematically focus on individual responsibility for staying 'healthy' while social factors such as education, health inequalities and environmental factors are systematically neglected. However, professional responsibility considering high standards of moral ethos of health professions does not necessarily buy into this logic. Instead, professionals who believe in prevention can also claim for interventions such as: free public access to sport activities, strict regulation of labels for unhealthy food, financial support for class- and age-stratified access to complex health information and, finally, high investments in public education, as educational level has been identified as a protective level against dementia, too. This would require an in-depth analysis of the existing gap between medical care and public health. Prevention – as primary prevention – and its normative framing do not fit very well into the common medical-ethical framework of self-determination, well-being, non-maleficence and justice. Public health practice differs significantly from medical practices because it focuses on the health of a population, while medial practice targets the illness of the individual. Public health requires explicitly the consideration of rather broad social settings of health conditions (e.g. considering different social groups, education, working and living conditions). Furthermore, public health ethics expands the space- and time-dimensions: population is thought of as a forward-facing entity, and timescales often cover a whole generation. The professional norm of 'preventing', in contrast to 'curing', targets diverse social groups and different underlying concepts of illness and health. Promoting too simplistic messages – even with a naive assumption that this does not harm – can, however, lead to blaming, shaming and unjustified responsibilization of the individual who will then become many years later a person with dementia. This can become a realistic risk for those living with the illness. Problematically, the considerably long period between promoting prevention and recognizing the 'non-effects' goes hand-in-hand with blurring and forgetting about any professional wrong-doing. Professionals are unlikely to be criticized 20 years later that they had in the past promoted an ineffective idea, but patients might suffer from a particular new climate of blaming. Therefore, professional responsibility should entail a critical reflection about the general implications of prevention as well as a responsible statement that they will be held accountable. If promoting prevention on the basis of non-harm or general well-being can in the long run lead to serious harm for the patients, then it is a professional responsibility to be more cautious about the assumption of what is non-harming.

Conclusions

The three provided examples illustrate that professional *responsibility* with regards to health risk communication is quite complex and goes beyond the topic of genetic predictive testing or traditional 'informed consent'. Using the different examples I wanted to illustrate that professional risk communication requires a detailed understanding of the scientific facts behind risk calculation, an understanding of lay numeric and risk illiteracy, but also, particularly, a detailed understanding and self-reflection of the implied ethical norms and consequences, that expand on risk information. Risk information provided by health professionals, independent of whether this happens in an inter-individual setting or in a public health promotion context, should encourage patients to consider not only the right numbers but also what the ethical norms are behind the intention to inform patients and the public: is it to promote the right to self-determination including the right to not know, or is it rather the principle of well-being or even self-care that motivates the professionals for risk information?

The epistemic basis of risk information as such is not only at stake, but also the epistemic-causal dimension of the actions that can result from such risk information. Screening to detect early stages of a serious disease that can be treated versus the early stages of disease that cannot effectively be treated makes a tremendous moral difference. Bearing the 'consequences' of risk decisions should be, however, mainly on the patient's side. Patients can be socially sanctioned by self- or social blaming when they had not cared about risk information. This would be a morally unacceptable outcome and professional communication strategies ought to consider this problematic outcome systematically. Patients can also be physically harmed by not caring about the risk. This can be an outcome that results from the individual's wish not to know, but also from the multi-factorial reasons of such complex diseases. Therefore, not knowing is rather another factor in such multi-factorial events, than the only, causal one. This must be clearly stated within professional communication, too.

Disclosure statement

There are no financial interests or benefits arising from the direct applications of the author's research.

Acknowledgements

This work has been supported by a grant to the project *Mind the Risk* from The Swedish Foundation for Humanities and Social Sciences (Grant No:PR2013-0123).

Notes

1 www.cancer.gov/about-cancer/causes-prevention/genetics/brca-fact-sheet#what-are-brca1-and-brca2 [accessed: 22-1-20].
2 See the German website of the cancer foundation: www.krebshilfe.de/informieren/ueber-krebs/krebsfrueherkennung/ [accessed: 24-1-20]. It states as first sentence when opening the website: '*About 40 percent of all cancer could be avoided, so experts estimate, if people would live a more healthy life*' [own translation].
3 E.g. www.krebshilfe.de/informieren/ueber-krebs/krebsfrueherkennung/ [accessed 24-1-20]; www.cancer.gov/types/breast [accessed: 24-1-20], www.nhs.uk/conditions/breast-cancer/ [accessed: 24-1-20]; www.informedhealth.org/breast-cancer.2276.en.html [accessed: 24-1-20].

References

Ayo, N. (2012) Understanding health promotion in a neoliberal climate and the making of health-conscious citizens. *Critical Public Health* 22(1): 99–105.

Baier, K. (1991) Types of Responsibility, In: *The spectrum of responsibility*, ed. P.A. French, 117–122. New York: St. Martin's Press.

Bayertz, K. (1995) Eine kurze Geschichte der Herkunft der Verantwortung, In: *Verantwortung. Prinzip oder Problem?*, ed. K. Bayertz, 1–72. Darmstadt: Wissenschaftliche Buchgesellschaft.

Beauchamp, T.L., and J.F. Childress (2008) *Principles of Biomedical Ethics*. 6th ed. Oxford: Oxford University Press.

Beier, K., I. Jordan, C. Wiesemann, and S. Schicktanz (2016) Understanding collective agency in bioethics. *Medicine, Health Care and Philosophy*: doi: 10.1007/s11019-016-9695-4.

Berg, J.W., P.S. Appelbaum, C.W. Lidz, and A. Meisel (2001) *Informed consent: Legal theory and clinical practice*. 2nd ed. New York: Oxford University Press.

Brasure, M., P. Desai, H. Davila, V.A. Nelson, C. Calvert, E. Jutkowitz, M. Butler et al. (2018) Physical Activity Interventions in Preventing Cognitive Decline and Alzheimer-Type Dementia: A Systematic Review. *Annals of Internal Medicine* 168(1): 30–8. https://doi.org/10.7326/M17-1528.

Chadwick, R., M. Levitt, and D. Shickle (eds.) (1997) *The right to know and the right not to know*. Aldershot: Avebury.

de Jonge, E.T., M.J. Vlasselaer, G. Van de Putte, and J.-C. Schobbens (2009) The construct of breast cancer risk perception: need for a better risk communication?. *Facts Views Vis Obgyn* 1(2): 122–9.

Elmore, J.G., and G. Gigerenzer (2005) Benign Breast Disease – The Risks of Communicating Risk. *New England Journal* 353(3): 297–9.

Featherstone, K., A. Bharadwaj, and A. Clarke (2006) *Risky relations: Family, kinship and the new genetics*. Oxford: Berg.

French, P.A. (1991) Types and principles of responsibility, In: *The spectrum of responsibility*, ed. P.A. French, 113–16. New York: St. Martin's Press.

Gigerenzer, G. (2014) Breast cancer screening pamphlets mislead women. *BMJ* Apr 25; 348: g2636. doi: 10.1136/bmj.g263.

Hallowell, N. (1999) Doing the right thing: Genetic risk and responsibility. *Sociology of Health & Illness* 21(5): 597–621.

Hampel, H., S. Lista, St.J. Teipel, F. Garaci, R. Nisticò, K. Blennow, H. Zetterberg, et al. (2014) Perspective on future role of biological markers in clinical therapy trials of Alzheimer's disease: a long-range point of view beyond 2020. *Biochemical Pharmacology* 88(4): 426–49. https://doi.org/10.1016/j.bcp.2013.11.009.

Hansson, S.O. (2010) Risk: objective or subjective, facts or values. *Journal of Risk Research* 13(2): 231–8.

Hart, H.L.A. (1968) Legal responsibility and excuses, In: *Punishment and Responsibility: Essays in the Philosophy of Law*, 28–53. Oxford: Clarendon Press.

Hessling, A., and S. Schicktanz (2012) What German experts expect from individualised medicine: Problems of uncertainty and future complication in physician-patient-interaction. *Clinical Ethics* 7(2): 86–93.

Holl, J., H. Lenk, and M. Maring (2001) Verantwortung, In: *Historisches Wörterbuch der Philosophie*, 566–670. Basel: Schwabe.

Jonas, H. (1984) *Das Prinzip Verantwortung*. Frankfurt am Main: Suhrkamp.

Kihlbom, U. (2008) Autonomy and negatively informed consent. *Journal of Medical Ethics* 34(3):146–9.

Leefmann, J., M. Schaper, and S. Schicktanz (2017) The concept of 'genetic responsibility' and its meanings: a systematic review of qualitative medical sociology literature. *Frontiers in Sociology* 1(18). doi: 10.3389/fsoc.2016.00018.

Leibing, A., and S. Schicktanz (eds.) (2020) *Preventing Dementia? Critical perspectives on a new paradigm of preparing for old age*. Berghahn: New York.

Lemke, T. (2006) Genetic responsibility and neo-liberal governmentality: Medical diagnosis as moral technology, In: *Michel Foucault and power today. International multidisciplinary studies in the history of the present*, eds. A. Beaulieu and D. Gabbard, 83–91. Lanham, MD: Lexington Books.

Liede, A., M. Cai, T. Fidler Crouter, D. Niepel, F. Callaghan, D. Gareth Evans (2018) Risk-reducing mastectomy rates in the US: a closer examination of the Angelina Jolie effect. *Breast Cancer Res Treat* 171(2): 435–42. doi: 10.1007/s10549-018-4824-9.

Livingston, G., A. Sommerlad, V. Orgeta, et al. (2017) Dementia prevention, intervention, and care. *Lancet* 390 (10113): 2673–734. https://doi.org/10.1016/S0140-6736(17)31363-6

Nelson, H.D., R. Fu, K. Goddard, J. Priest Mitchell, L. Okinaka-Hu, M. Pappas, and B. Zakher (2013) Risk Assessment, Genetic Counseling, and Genetic Testing for BRCA-Related Cancer, Systematic Review to Update the U.S. Preventive Services Task Force Recommendation. *Evidence Syntheses, No. 101,* Rockville (MD): Agency for Healthcare Research and Quality (US); 2013 Dec. Report No.: 12–05164-EF-1.

Perry, J., S. Wöhlke, A.C. Heßling, and S. Schicktanz (2017) Why take part in personalised cancer research? Patients' genetic misconception, genetic responsibility and incomprehension of stratification—an empirical-ethical examination. *European Journal of Cancer Care*, 26(5): https://doi.org/10.1111/ecc.12563.

Potter, Van R. (1970) The Science of Survival Perspectives in Biology and Medicine. *Bioethics*, 14(1), 127–53.

Raz, A., and S. Schicktanz (2016) *Comparative Empirical Bioethics: Dilemmas of Genetic Testing and Euthanasia in Israel and Germany*. Basel: Springer.

Schicktanz, S. (2017) The visionary shaping of dementia research: Imaginations and scenarios in biopolitical narratives and ethical reflections, In: *Planning Later Life. Bioethics and Public Health in Ageing Societies*, eds. M. Schweda, L. Pfaller, K. Bauer, F. Adloff, S. Schicktanz, 205–27. London: Routledge.

Schicktanz, S. (2018) Genetic risk and responsibility: reflections on a complex relationship. *Journal of Risk Research* 21: https://doi.org/10.1080/13669877.2016.1223157.

Schicktanz, S. (2020) If dementia prevention is the answer – what was the question? Observations from the German Alzheimer's disease debate, In: *Preventing Dementia? Critical perspectives on a new paradigm of preparing for old age*, eds. A. Leibing and S. Schicktanz. Berghahn: New York.

Schicktanz, S., and F. Kogel (2014) Genetic responsibility revisited: moral and cultural implications of genetic prediction of Alzheimer's disease, In: *Genetics as social practice*, eds. B. Prainsack, S. Schicktanz, G. Werner-Felmeyer, 199–218. Farnham: Ashgate. Reprint, Routledge.

Schicktanz, S., and M. Schweda (2012) The diversity of responsibility: The value of explication and pluralization. *Medicine Studies* 3: 131–45.

Schicktanz, S., M. Schweda, J. Ballenger, P. Fox et al. (2014) Before it is too late: Professional responsibilities in late-onset Alzheimer's research and presymptomatic prediction, *Frontiers in Human Neurosciences* 8: 921.

Schweda, M., and L. Pfaller (2020) Responsibilization of aging? An ethical analysis of the moral economy of prevention, In: eds. A. Leibing and S. Schicktanz, *Preventing Dementia? Critical perspectives on a new paradigm of preparing for old age*. Berghahn: New York.

Strawson, P. (2008) *Freedom and resentment and other essays*. London: Routledge.

Ter Meulen, R., and F. Jotterand (2008) Individual responsibility and solidarity in European health care: Further down the road to two-tier system of health care. *Journal of Medicine and Philosophy* 33: 191–7.

Waller, J., E. Douglas, K.L. Whitaker and J. Ward (2013) Women's responses to information about overdiagnosis in the UK breast cancer screening programme: a qualitative study. *BMJOpen* 3: e002703. doi: 10.1136/bmjopen-2013-002703.

Whitehouse, P.J. (2003) The Rebirth of Bioethics: Extending the Original Formulations of Van Rensselaer Potter. *The American Journal of Bioethics* 3(4): 26–31.

Wöhlke, S., and J. Perry. (2019) Responsibility in dealing with genetic risk information. *Social Theory & Health*: https://link.springer.com/article/10.1057%2Fs41285-019-00127-8#Sec2.

Wöhlke, S., J. Perry, and S. Schicktanz (2018) Physicians' communication patterns for motivating colorectal cancer patients to biomarker research: empirical insights and ethical issues. *Clinical Ethics*: https://doi.org/10.1177/1477750918779304.

World Health Organization (WHO) (2019) Risk reduction of cognitive decline and dementia. WHO guidelines. Geneva.

Yoder, S.D. (2002) Individual responsibility for health: Decision, not discovery. *Hastings Center Report* 32: 22–31.

Young, I.M. (2011) *Responsibility for justice*. Oxford: Oxford University Press.

Zielonke, N., A. Gini, E.E.L. Jansen, et al. (2020) EAM; EU-TOPIA consortium 2020. Evidence for reducing cancer-specific mortality due to screening for breast cancer in Europe: A systematic review. *European Journal of Cancer*: doi: 10.1016/j.ejca.2019.12.010. [Epub ahead of print]

7 Genetic testing and risk perception in the context of personalized medicine

How familial relationships matter?

Sabine Wöhlke, Marie Falahee and Katharina Beier

Introduction

Genetic information is of increasing importance in medical research and health care. For example, progress in predictive medicine enables the early detection and diagnosis of diseases. On the one hand, genetic knowledge may improve the provision of health care services, e.g. by determining the most effective therapy and avoiding ineffective or even futile treatments. Moreover, genetic information can have beneficial effects for the individual, e.g. by revealing health-relevant information for the prevention of certain diseases (Burke 2014). On the other hand, however, genetic information can also have detrimental effects. For example, knowing that one has an increased or even certain risk of developing an untreatable disease may not only be highly burdensome but can also increase the risk of psychological stress, discrimination and stigmatization (Kalokairinou et al. 2018; Ross et al. 2013; Borry et al. 2014).

Given these ambivalences, a person's right to autonomy plays a crucial role in the handling of genetic information. While in principle, autonomy is a guiding principle for all kinds of health-related information, there is special emphasis on this in the genetic context. Many legislations and ethical frameworks grant individuals a 'right to know' or 'right not to know' and stress the importance of genetic privacy (Gilbar et al. 2016; Etchegary and Fowler 2008; Andorno 2004). In this context, genetic information is typically presented as being specific to the individual – a unique individual 'fingerprint' that distinguishes it from others. However, this perception neglects the supra-individual dimension of genetic risk information. In fact, knowing about an individual's genetic disposition may also reveal something about potential health risks for family members (Gaff et al. 2005). This is even more the case with genome-based testing as this may 'have far greater implications for family relations than does directed, conventional genetic testing' (Mendes et al. 2017: 3). For this reason, debates about disclosing or not disclosing genetic information to family members (or testing or not testing on behalf of family members) do not sufficiently reflect the implications for family dynamics. In fact, framing these questions in terms of individual autonomy ignores the fact that *any* decision taken will impact familial relationships and dynamics (Gilbar 2007; Schicktanz 2018; Wöhlke and Perry 2019).

In this chapter, we analyse the role and implications of familial relationships and responsibilities for the handling of genetic information. Thereby, we restrict our analysis to genetic testing in the context of personalized medicine in clinical contexts and exclude related fields such as prenatal genetic testing. We aim to show how familial genetic interrelatedness transcends the logic of individual genetic decision-making. In particular, familial interdependencies challenge individualistic notions such as the right (not) to know and call for a more nuanced reflection on familial responsibilities and obligations. Our analysis is based, first, on an evaluation of theoretical analyses as well as pertinent empirical studies dealing with the familial implications of genetic testing. Next, we provide two empirical case studies in order to illustrate the complex, often implicit, negotiations around genetic testing and risk disclosure amongst family members. Against this background, we finally reflect on different approaches towards reconciling individual with collective (i.e. family) interests, and on their ethical and practical implications.

Family and family responsibilities

The family is a highly relevant entity in medicine. Health care decisions have potential effects on 'family goods', for example established patterns of responsibility-relations or strong emotional dependencies (Kihlbom and Munthe 2017). In medical practice, family members serve as living organ donors, act as proxies for incapacitated relatives, and often directly care for seriously ill or elderly family members. Despite this, the family is rarely the focus of bioethical analyses (Beier et al. 2016; Verkerk et al. 2015; Atkinson et al. 2013). Given the importance of individual autonomy in modern medicine, it may seem challenging to take the interests of family members into account without compromising individual autonomy. This is directly relevant for the context of genetic testing and the handling of genetic risk information within the family. Before we discuss this in more detail, we will briefly reflect on what a family is and highlight the nature of familial commitments and responsibilities. This reflection is based on the premise that the availability of genetic information creates new familial responsibilities.

Intuitively, we all know what a 'family' is and treat the term as self-evident. However, it is not clear that people necessarily mean the same thing when speaking of 'the' family. For example, for some people or in certain contexts the family is understood as a biological-genetic entity.

From this perspective, family membership is seen as something objective. Others regard the nature of the relationship as constitutive of family membership. From this perspective, being a member of a family implies, amongst other things, lasting bonds, mutual affection, intimacy, care, commitment and support (Wiesemann 2010). According to Iris Marion Young's influential account, family is defined by 'people who live together and/or share resources necessary to the means of life and comfort; who are committed to taking care of one another's physical and emotional needs to the best of their ability; who conceive themselves in a relatively long-term, if not permanent relationship and who recognize themselves as family' (Young 1997: 196). This emphasis on the subjective self-perception of those involved in familial relationships (also referred to as 'doing family') is particularly relevant for alternative family models, e.g. lesbian couples or single mothers by choice.

Debates about the nature of the family are often linked to the question of whether special familial responsibilities exist. Two positions can be roughly distinguished here: voluntarists deny this by stressing that only consented-to relationships can ground special obligations, while non-voluntarists argue that special obligations can also arise from non-voluntary relationships (Scheffler 1997; Honneth and Rössler 2008). If one follows this latter position, the question remains how the moral authority of familial responsibilities can be explained and justified and what they mean in practice. Special responsibilities within personal relationships may be accepted for different reasons, e.g. due to a certain type of interaction, shared group membership or the inherent normativity of the relationship itself (Scheffler 1997). In fact, if it comes to families, different aspects may evoke senses of responsibility amongst its members. As Rosamunde Rhodes points out,

> neither DNA nor social history fully explains our ethical responsibilities to one another. (...) More likely reasons would be related to the intimacy and dependency of our previous relationship, or the strength of our feelings, or the history of our interactions, or something about our relative wherewithal and neediness. (...) The bonds that have moral weight and give us thick responsibilities to one another typically include a social component.
>
> (Rhodes 1998: 21)

Those who accept special responsibilities amongst family members often point out that these responsibilities imply more than just non-interference but cooperation and support (Smith 2010; Crouch and Elliot 1999). For example, Juengst characterizes the ethics of family life[1] in terms of three value commitments: loyalty, intimacy and security (Juengst 1999). Loyalty means that family members actively promote each other's well-being by providing mutual assistance, perhaps even setting their individual interests aside for the sake of other family members. Intimacy implies, amongst others, the readiness to share information with family members that one would not reveal in other contexts.[2] Finally, Juengst uses the term 'security' to describe the obligation of family members to protect and defend each other (Juengst 1999). This obligation may not least derive from the fact that family members are in a unique position to help each other (Rhodes 1998: 24).

However, there may be different perceptions of the *exact scope* of familial obligations, even amongst the members of a single family. If we accept that the quality of a relationship and the value that we assign to it defines the scope and content of our familial responsibilities (Scheffler 1997; Jeske 1998; Schweikard 2015),[3] it becomes clear why our senses of familial obligation differ. Moreover, what is owed to other family members is rarely made explicit, except during familial controversies. Most of the time, individuals have a rather implicit sense of their familial responsibilities, whose practical implications, however, become visible in daily familial interactions. In the next sections, we will provide an analysis of the challenges that genetic testing implies for family relationships and responsibilities. We will do this by looking at the results of previous theoretical and empirical studies.

Genetic testing and the family

Genetic illness within the family has been well described in the literature (Featherstone et al. 2006; Gaff et al. 2007). Active decision-making regarding testing, coping with the knowledge of genetic risk information, and its incorporation into personal and family life involves fundamental challenges (Mendes et al. 2017). According to Mendes et al. the negotiation of genetic information depends on four essential, interrelated aspects: the probability of development of the genetic disease, the clinical severity, the age of onset, and the availability of treatment or preventive measures (ibid.). In the context of predictive genetic testing, it is generally assumed that it is the responsibility of the individual undergoing testing to disclose the results to her relatives (Daly et al. 2016). This is typically justified by the argument that family members have the 'right to know' in order to be able to make their own autonomous decisions about genetic risks. Autonomy is seen as the central ethical principle in predictive genetic testing, but it also entails moral obligations to disclose and communicate these risks (Arribas-Ayllon et al. 2011). However, family dynamics that include individual patterns of rules and boundaries (e.g. roles of responsibility) influence preferences for sharing and wanting to know about health information (Mendes et al. 2017).

Mostly strikingly, developments in genetic testing are likely to exacerbate potential conflicts or at least tensions between family members. Whereas in former times family records were needed, as well as the informed consent of other family members, today genetic tests can be taken without even informing one's relatives (Juth 2014). Therefore, the importance of kinship in terms of common ancestry, shared genetic material and shared risk is increasing (Featherstone et al. 2006). However, family dynamics that include individual patterns of rules and boundaries (e.g. roles of responsibilities) influence preferences for sharing and knowing health information with each other (Mendes et al. 2017). Moreover, there are different constellations of responsibilities within families that may not least be influenced by gender aspects. For example, Hallowell's studies have shown that women are ready to get tested even if this means that they have to give up their own right of not knowing (Hallowell 1999; Hallowell 2003). In fact, it has been pointed out that testing children or other family members can breach the confidentiality and the privacy of genetic information and ignore the specific moral status of the person (Borry et al. 2014).

Decisions for predictive genetic information can have far-reaching consequences for family members, for example, when an individual decides to be tested and acquires information that may be relevant for family members. Or perhaps a relative who suffers from a disease and is seeking clarity about their genetic condition requires family members to be tested (if a family member is tested and it is confirmed that they have a certain mutation ('index person'), it is possible to deduce with a certain probability whether relatives also have this mutation) (Juth 2014). In many cultures and communities, the involvement of the family in decision-making in genomics research will be taken for granted due to shared understandings of the identity and obligations of the individual (Meslin and Juengst 2019). These could potentially lead to clashes between individual interests and family interests, but genetic testing could also have far-reaching implications for family members even if there is no conflict. In fact, there is evidence that conflicts may be more of a theoretical than an empirical problem; in practice, family members seldom explicitly refuse to disclose a genetic risk to relatives. However, there are cases of what might be called 'passive non-disclosure', i.e. family members want to inform relatives but for various reasons fail to do so (Mendes et al. 2017; Leonard and Newson 2010). The immediate conflict is then not carried out within the family (Mendes et al. 2016); rather, negotiations about one's role and responsibility towards family members, and how to deal with test results and already acquired genetic knowledge, are carried out alone. The hesitation to disclose something (passive non-disclosure) might be caused by feelings of guilt towards those who are tested positive, or those who receive unfavourable results may blame their parents. The emotional closeness of a relationship may also play a role, whereby an emotionally intimate relationship may promote greater willingness to disclose information. A problematic or conflictual relationship, or a lack of contact, may prevent the disclosure of information (Gilbar 2007). Further, a lack of knowledge about the importance of genetic information could influence the disclosure. Other studies have shown that the desire to prevent harm to family members by protecting them from bad news can be a motive for not sharing information (Timmermans and Buchbinder 2010).

Donchin argues that test results may disrupt family expectations, impelling members to reshape their self-identities and renegotiate family positioning (Donchin 2000: 245). Several empirical studies have shown the complexity of dealing with genetic information within the family. For example, Wöhlke et al. (2020) show that people who have undergone genetic testing are generally willing to share genetic information with family members. However, there are cultural differences in the extent to which people felt obliged to pass on information to the family. For example, Germans tend to emphasize their individual rights, whereas Italians feel a much stronger obligation to pass on information to the family. However, in some cases communication about genetic information can be problematic. It has been shown that only 15–20 per cent of at-risk relatives receive information about this risk. Thereby, information may either be completely withheld or delayed or incomplete (Mendes et al. 2017: 4). The difficulties experienced by tested persons regarding the disclosure of risk information to family members is also illustrated by an international study which found that there is a quite strong pro-disclosure attitude amongst individuals, but many seem to be hesitant to disclose this information themselves. As many as 42–47 per cent of the patients were of the opinion that doctors should report results to family members even without their consent. Moreover, 33–42 per cent of patients believed that doctors should even approach relatives unsolicited to inform them (Wertz and Fletcher 2004). It became clear that the challenges of predictive genetic testing for family (relationships) and responsibilities are very complex. We will illustrate this in a deeper way in the next section by taking a closer look at the perspectives of healthy lay people on the one hand and the perspectives of affected families on the other.

Genetic decision-making within the family: public perspectives on management of genetic risk information

People are increasingly able to access digitized data about their health, ranging from information provided by self-tracking and fitness apps to electronic patient records (Wöhlke et al. 2020; Rexhepi et al. 2018). Genetic information has become widely available, presenting people in both health care and commercial contexts with a variety of possibilities regarding the application and utility of sharing such information. Studies show that public interest in genetic information is high (Townsend et al. 2012). Genetic tests can confirm or rule out a suspected genetic condition, or determine a person's chance of developing or passing on a genetic disorder, and in some cases, they provide relevant information that can be used to family members' benefit (Burke 2014; Mendes et al. 2015).

In an explorative qualitative study,[4] Wöhlke et al. (2019; 2020) examined the ethical aspects shaping risk perception in the context of genetic testing.[5] We aimed to explore the perspectives and opinions of members of the public about genetic testing and its implications to increase understanding of the explicit social and ethical issues involved. We conducted a focus group study which presented participants with fictional scenarios, describing different kinds of genetic testing.[6] The scenarios described the results of a genetic test to predict the risk of developing a serious disease in the future, such as breast cancer or Alzheimer's disease, or a genetic test to predict the likelihood that a person affected by colorectal cancer would benefit from neoadjuvant chemo- and radiotherapy. Predictive testing for individuals in clinical settings was positively perceived by participants, and associated with a positive potential impact on their own lives.

> I mean, perhaps also with regard to an advance health care directive or so. I mean that you might still be able to decide for yourself whether you perhaps want to move to a care home later or want to receive outpatient treatment. If you're no longer able to decide on your own at a later point that you could still make the decision before. [...]
>
> (Single, 18–25 years old, no children, previous experience with genetic testing; Wöhlke and Perry 2019: 11)

Predictive genetic testing was supported when a person had a family history of the disease, or when treatment existed for the disease tested, particularly in relation to breast cancer. In contrast, participants couldn't imagine that it would make more sense for patients to undergo cancer treatment without receiving chemotherapy (and therefore have more quality of life). Interestingly, these participants also expressed negative views about predictive testing for Alzheimer's disease. The value of predictive genetic testing was negotiated against the background of interrelated factors, including the perceived severity and treatability of a disease, the impact on future decision-making and planning as benefit to the individual and their family, and an individual's personal situation (especially age-negative viewpoints were associated with the concept of uncertainty). A test result predicting onset of Alzheimer's disease creates new and dramatic dependencies in the family, which had the effect that the participants were more likely to be against such a test:

> In the past, no one cared about it. That's just the way it was. People lived with their family, elderly people. They sometimes were a little scatter-brained, so what (laughs), that wasn't so bad, right? You just managed together. And today, everything's already so perfectly classified, everything. I don't know whether that's so good. Maybe you

> worry too much as a consequence. I think if it's severe forms that significantly change the behaviour, right? That's sometimes the case, especially when it comes to Alzheimer's disease. That changes the behaviour so fundamentally, then it's of course advisable to help the persons professionally, right?
>
> (Single, 51–70 years old, no children, no experience in genetic testing; Wöhlke and Perry 2019: 13)

Some participants argued that family members should have the right to receive genetic risk information in case of a hereditary genetic disease. Not sharing this important information with other family members could represent a great burden for the person being tested. However, participants also related that the individual should choose which family members they shared such information with:

> Okay, I'd tell my daughter, among others, for example, because she really should get herself tested too. Because it's really in our [...] I just counted, there were at least five cases of death, really within my blood-related family. But well, you can't generally, you have to decide for yourself [...] whether you, whom you'll tell.
>
> (Widowed, 70+ years old, three children, no experience with genetic testing; Wöhlke and Perry 2019: 14)

The decision of the individual therefore may not be consistent with the interests of all family members. Personal situation and age strongly influenced participants' perceptions of the impact and relevance of genetic risk information on life planning. Important considerations included whether relevant goals in life had been achieved, and the personal responsibilities of an individual. Accordingly, participants stated that to receive a high disease risk status as a young person (under 30 years of age) was burdensome. On the other hand, the exclusion of a disease through genetic testing was perceived as a form of great relief among middle-aged participants, especially when a disease ran in the family.

Family responsibility was rarely used as a justification to have predictive testing. Instead, participants often stressed that an individual should be guided by reflection on the personal implications and consequences of a genetic test result for themselves.

> Yes, I also think that's a completely individual decision. I mean, I don't think that there are any, like, general yardsticks. I mean, of course, if someone from my immediate family and personal circle received such a diagnosis now, I would want the person to tell me, but that's of course the decision of every individual [...].
>
> (Single, 18–25 years old, no children, no experience with genetic testing; Wöhlke and Perry 2019: 14)

Interestingly, treatable diseases were considered from the perspective of the individual, while non-treatable diseases were considered on a collective level, involving family, caregivers or society.

A genetic test was perceived positively in terms of potential impact on personal context (e.g. age, having children, caring responsibilities etc.), health care status and the need to take preventive lifestyle measures.

> Yes, I mean, someday, it's no longer just about yourself, you're rather on your own while still young. That your parents outlive you is so far from (laughs) reality for you at that age, but someday, you'll have a family, and someday, you'll have responsibility, and you'll also be part of a family you started yourself, and I think that really is a different situation then. […] I think it really does have something to do with age, maybe not necessarily only with the attitude of yours, which changes, but also with the situation you're in […].
>
> (Single, 26–35 years old, no children, no experience with genetic testing; Wöhlke et al. 2019: 5)

This quote illustrates that some participants feel more responsible towards their family members when it comes to the risk of diseases that might affect their relatives in later life.

Regarding the value of a genetic test for treatable or non-treatable diseases, there was a clear difference between the responses of younger and older participants. Older participants generally do not consider genetic testing to be so important, even when it comes to testing for a specific disease such as cancer. For them, preservation of quality of life was the most important consideration – something that could be facilitated by genetic testing. In contrast, younger participants viewed the life course as binary (life/death) and associated non-treatability of a disease with death.

In summary, beliefs about diseases and contextual personal and cultural factors affected participants' views about the utility of predictive genetic testing in the scenarios used. The complexity of notions of responsibility was highlighted in this context. Currently, in Germany, physicians may only prescribe a genetic test if there is a medical indication for the individual in their care (e.g., family history or symptoms of a genetic disease) (Wöhlke et al. 2019: 9). Narratives concerning dynamic individual and cultural conceptions of self and identity played a pivotal role in deliberations about the value of predictive genetic testing for different diseases.

With regards to the communication of genetic risk information, participants referred to both *individual* and *relational* conceptions of genetic risk. Concerns were formulated in terms of the self and others (relatives or society) to whom they felt they belonged, and/or whom they felt obliged to care for. These individual and family responsibilities were more important to participants than numerical risk estimates. Therefore, predictive genetic test results can be seen as a form of knowledge that should or should not be shared depending on an individual's conception of their relations with others. This creates new options for acting but also for power and dependency (e.g. asymmetries of knowledge between families).

These findings illustrate the complexity of public views on predictive genetic testing and indicate that decision-making in this context is undertaken in a relational context. A relational perspective, which refers to moral decisions and considers the interests of others, can be seen as an alternative to an individual approach to autonomy (Schicktanz 2018; Wöhlke and Perry 2019).

Perspective of affected people: managing risk of rheumatoid arthritis

The study described earlier addressed perspectives on predictive genetic testing relating to diseases that are widely perceived to have a severe negative impact on the lives of an individual and their relatives. Predictive genetic testing for these diseases is increasingly becoming integrated into clinical practice, however the range of diseases that this approach is appropriate for is likely to expand as understanding of disease aetiology and technological innovation increase. Rheumatoid arthritis (RA) is a debilitating chronic autoimmune disease of multifactorial aetiology characterized by painful and destructive inflammation of the joints, often associated with fatigue and depression. Effective treatments for RA are available for most patients, but they are usually required for the long term, and are associated with potentially serious adverse effects. There is therefore an increasing research focus on the identification of those at risk of developing RA to facilitate early treatment and preventive intervention (van Boheemen and van Schaardenburg 2019). It is therefore increasingly likely that management of RA will shift towards prevention, rather than treatment of symptoms. Understanding the perceptions of those likely to be affected by predictive and preventive strategies is key for developing efficient future clinical practice and appropriate support for patients. Furthermore, tests providing information about the risk of common chronic conditions, such as RA, are increasingly available directly to consumers outside of clinical settings. Yet an understanding of how commercial provision of risk information should be regulated, or the resources needed to support consumers to interpret and manage such information, is limited. Within the *Mind the Risk* project, we have conducted a range of systematic reviews and qualitative interview studies to explore the perceptions of a range of stakeholders about predictive testing for chronic inflammatory conditions (Mosor et al. 2019; Falahee et al. 2016a; Bayliss et al. 2016), including patients with a diagnosis of RA and their first-degree relatives about assessment and communication of RA risk (Falahee et al. 2016b; Stack et al. 2016; Falahee et al. 2019). Quotations from several of these studies are used in the remainder of this section to illustrate ethical issues associated with perceptions of risk and the value of predictive testing in this context, and how these were shaped by family relationships.

Perceptions of disease risk are a critical determinant of health behaviour (Ferrer and Klein 2015). However, disease risk perceptions are shaped by numerous factors beyond the objective estimates of disease probability provided by health care professionals. An individual's perception of risk is often influenced more by personal experiences, theories of inheritance, and subjective illness perceptions than by numerical estimates of disease risk (McAllister 2003; Walter and Emery 2005; Lautenbach et al. 2013). For example, relatives may feel that their risk of developing a disease is higher if they resemble an affected relative in other ways:

> I seem to follow my mum in absolutely everything, like my brother and sister they're quite like my dad, they never get ill, they never catch a cold. Whereas if there's a cold going around I will get it and the same with my mum [...] So I was a bit like 'oh, maybe I'll get it [RA]'.
> (First-degree relative of RA patient; Stack et al. 2016: 5)

The degree of experience that relatives have with a disease is likely to affect perceptions of disease impact and susceptibility. Our studies of family members with RA illustrate the extent to which such experience varies within and between families. In some cases, relatives' exposure to the condition may be at least partly managed by the patient with the relevant condition. Some relatives in our study described their parents' reticence to talk about their RA, and others felt that their parents tried to 'protect' them from concern about RA:

> I never had that information of what happens, how you're made at higher risk, I've never had that in like black and white [...] which makes me think she doesn't know or maybe she's just trying to protect me like a mother does. Because I think she was quite worried about me taking part [...] she's quite worried about what I'd find out.
> (First-degree relative of RA patient; Stack et al. 2016: 6)

Perceptions of risk are also likely to be influenced by common misperceptions about the causes and severity of RA (Simons et al. 2016, 2017), and both patients and relatives in our studies had incomplete or inaccurate knowledge about risk factors for RA. Patients also related that even family members in close proximity to them do not fully comprehend the severe negative impact that RA has on their life, and that this was likely to be a major barrier to engaging at-risk family members with risk reduction approaches (Falahee et al. 2016b).

Our studies have further illustrated how individuals often conceptualized risk information in binary terms – confirmation that a person will either develop the disease or they will not, often relating to the experiences of other family members. This conceptualization was linked with positive views about predictive testing for RA as a method to rule in or out the disease:

> So, if there is a link with, well, mum having it and me having it, really if that flagged up they could have done that test right at the start and say yes, it is or no it isn't.
>
> (RA patient; Falahee et al. 2016b: 1563)

On the other hand, understanding of the inherent uncertainty associated with risk information for multifactorial conditions, was associated with negative views of predictive testing and preventive intervention in this context:

> That's really difficult, because you know that the tests show that you're at a higher risk of getting it, but that's not saying that you are going to get it, and so are you taking medication unnecessarily?
>
> (RA patient; Falahee et al. 2016b: 1563)

Where predictive tests are unable to rule in, or rule out, future RA and preventive interventions are unable to guarantee that an individual will not go on to develop the disease, patients may be less than eager to encourage their relatives to engage with such approaches, and those at risk may be unlikely to accept them. Risk counselling initiatives to support preventive strategies can therefore usefully convey the probabilities of both positive and negative outcomes to facilitate informed decision-making in this context, and personalized approaches can positively impact on risk-related health behaviours (Sparks et al. 2018).

An important concern across our qualitative studies relating to RA risk was the potential for information about disease risk to cause anxiety. This related not only to increased anxiety for those identified as having an elevated risk status, but also for other family members. Relatives described not wanting to engage with predictive testing in order to prevent their parents from being anxious about their well-being:

> I wouldn't want my mum to worry that I was going for this test [...] to know that if in five years' time I'll get it, I don't want her to know that because I think that would worry her more than anything.
>
> (First-degree relative of RA patient; Stack et al. 2016: 7)

Others described concerns about the implications of risk information about oneself for other family members:

> You only worry too much and rack your brain, because then I have to consider that my children could get it too and then you would worry too much. It's more comfortable to avoid it.
>
> (First-degree relative of RA patient; Stack et al. 2016: 6)

Whilst there is initial evidence that recipients of information about RA risk do not experience increased distress, current understanding of the ways in which disease risk information impacts on others, and is shared within and beyond family networks, is limited. Such knowledge will be essential to guide policy and practice in this area.

Whereas a few patients in our study described feeling a responsibility to share information about RA risk to their relatives, decisions about risk communication within families were not simply based on notions of relatives' right to know, or patients' duty to share risk information with family members. Instead, stories about family communication about RA risk involved a complex selection process, involving careful assessment of individual relatives' receptivity to such information and their capacity to act on it:

> I think it depends on the people themselves really. Some people want to know about it, and some don't [...] I think you've got to know which person you can talk to about it.
>
> (RA patient; Falahee et al. 2016b: 1560)

In some cases, individual family members were excluded from information shared within otherwise close-knit family units:

> I talk openly with all my family about it. They're aware. No, I've got no qualms about that. I don't think it's anything to be ashamed of. It's just unfortunate, isn't it, if you've got it. [Interviewer: There are some members that you wouldn't talk to about it?] Well, just my sister.
>
> (RA patient; Falahee et al. 2016b: 1561)

In other cases, more pragmatic considerations determined the appropriateness of discussing RA risk with other family members, such as other diseases that a relative may be dealing with, or how much time they have available to them:

> Oh, they're busy women in their forties and they can't be bothered to take the time out to do things like that.
>
> (RA patient; Falahee et al. 2016b: 1561)

These findings illustrate the complexity of RA risk communication within families and are consistent with those described earlier in other disease settings. In summary, perceptions of risk are highly subjective and influenced by intra- and inter-familial variables. Reactions to risk information extend beyond the individual to include consideration of impact on other family members. Even within close-knit families, patterns of disclosure may not be predicted by notions of responsibility, or objective potential for family members to benefit from knowledge about their personal risk status.

The development of ethical frameworks to guide management of risk information in clinical practice that can incorporate this complexity is a challenge that becomes increasingly relevant as predictive algorithms continue to improve, and the current impetus towards the development of preventive approaches to common diseases such as RA becomes established.

Discussion: handling genetic risk information within the family

When it comes to the handling of genetic risk information within the family, ethical analyses are often concerned with the question of whether there is a duty to disclose or not and – correspondingly – whether there is an absolute right to know or not to know. Answers to these questions are usually provided in a differentiated manner. For example, Rhodes argues that limitations to genetic ignorance would be morally legitimate if the genetic information could be expected to make a difference to the decision-making of another family member (Rhodes 1998). Others consider disclosure to be warranted if there is a reasonable therapy for the condition in question available, and if the sharing of genetic information is likely to avert serious harm from other family members (Andorno 2004: 437; Juth 2014). However, some unknown factors remain. For example, it cannot be taken for granted that relatives would automatically wish to have this – often insecure – knowledge about their genetic risk. Despite these differentiated reflections on disclosure/non-disclosure, there has been criticism that the ethical discourse is based on an 'internalized normative expectation' for a duty to disclose genetic information within the family. According to Arribas-Ayllon et al., 'the dominant and normative explanations for this non-disclosure are in fact Trojan horses, they appear sympathetic to family difficulties while at the same reinstating a deficit of family communication. Families that cannot or will not engage in risk communication are deemed hazardous, incompetent or in denial' (Arribas-Ayllon et al. 2011: 20).

A truly differentiated picture of familial decision-making requires openness to the complex dynamics of familial decision-making and the variety of factors – personal and inter-relational – that impact decisions on genetic testing and risk communication within the family. Even though there is broad readiness amongst tested persons to share genetic information with their kin, in practice this sense of familial responsibility implies quite differentiated reflections. As our review of empirical studies and the two presented case studies show, timing, the severity of the condition, the impact that the information might have on other family members, the type/quality of the relationship, as well as assessments of specific family member's ability to cope with this knowledge are considered (Clarke et al. 2005; Atkinson et al. 2013). In practice, this can support decisions for both disclosure and non-disclosure.

Obviously, there is a difference between reflections that underpin decisions to undergo genetic testing and decisions about disclosure/non-disclosure. In the former, decisions seem to be less inspired by familial responsibility than by reflections about the potential impact of this step on one's own life; in the latter, notions of familial responsibility are inevitably involved if the individual is confronted with hereditary genetic risk information. Even if there is no open, explicit communication about genetic risk disposition within the family, there is empirical evidence that individual decision-making about handling genetic test results is strongly influenced by relational concerns in this context. Rather than focusing on individual autonomy and the related rights to know or not to know, decisions are taken in consideration of the interests of individual family members but also the family as whole. This may also help to explain the phenomenon of passive non-disclosure. Awareness of these complex relational dynamics may lead individuals to postpone the decision of whether or not to share genetic information with family members.

Outlook: reconciling individual and familial interests in medical practice

Proponents of relational autonomy have argued that 'both communal matrices and individual decision-making capacities' should be respected in the context of genetic decision-making (Donchin 2000: 247). How this can be realized in practice, without sacrificing the individual's point of view to an outright collective perspective or vice versa, is still an open question. In the ethical literature, different approaches to dealing with the interrelatedness of genetic information can be found. According to the model of genetic unity, familial agreement is required for the genetic service to share genetic information with family members (unless the individual that undergoes testing wishes to inform her relatives herself) (Leonard and Newson 2010: 203). Practically, however, this model seems to imply that in the absence of familial agreement an individual could be prevented from undergoing genetic testing at all. This seems close to a sacrifice of individual autonomy rather than a balancing with communal concerns. The comity model, as another approach, aims to balance individual autonomy and notions of relational autonomy by including a sense of familial responsibility. Comity implies 'considerate behaviour towards others'. The model is meant to be reflective of the existing practice 'that genetic practitioners already encourage clients to disclose, as well as [...] that family members consider a sense of responsibility when deciding to disclose' (Leonard and Newson 2010: 205). While the comity model builds on notions of genetic solidarity and altruism amongst family members, decisions against disclosure would still be reconcilable with familial responsibility.

Based on a review of studies, Gilbar points out that there exists a sense of disharmony: 'On the one hand, patients value their legal and ethical right to confidentiality, rejecting the possibility that doctors may disclose medical information to relatives without their consent. On the other hand, they have a strong sense of responsibility to their relatives, and are willing to share information with them and to justify doctors' disclosure without consent even when physical harm to them cannot necessarily be prevented' (Gilbar 2007: 691). For Gilbar, this ambivalence explains the simultaneous independence and dependence of people – from the perspective of relational autonomy, the wish to control one's genetic information and the interest in sharing it are not in contradiction (ibid.)

However, leaving this decision with the tested individual alone may be too much of a burden. For this reason, Leonard and Newson propose a partnership model between physicians and individuals in order to deal with genetic information within the family. According to this approach, at-risk family members are identified along with the person who undergoes testing. In addition, the tested person is asked whether she is comfortable with contacting these at-risk family members on her own, or if she wants the genetic service/physician to approach them. If the tested person sees barriers for communication (e.g. anxiety of a family member), the genetic counsellor/physician will help the person to decide whether the relative would benefit from knowing. Even though the partnership model seems to encourage disclosure in general, it remains sensitive to the individual's interest in controlling her genetic information as well as to intimate dynamics of families and familial responsibilities, which must be brought into balance on a case-by-case basis.

Notes

1 Juengst's focus is expressly on American family life. However, the virtues he points out seem to be valid for other Western societies, too.

2 For this openness, mutual trust relationships amongst family members play a crucial role (Beier et al. 2016).

3 In response to the inherent problem of non-voluntarist accounts that 'special responsibilities could be rendered void by the relative conditions of the relationship' (Farmer 2010: 46), special familial responsibilities have been conceptualized as 'prima facie duties' that can supersede, but not void, special obligations in the case of undesirable conditions.

4 For further information see http://134.76.20.153/egm/index.php?id=146&L=1&L=1

5 Our German sub-project in *Mind the Risk* examines ethical issues regarding the communication of genetic risks, the perception of responsibility in the context of genetic diagnostics, as well as lay people's attitudes towards the possibilities of genetic diagnostics and direct-to-consumer genetic testing. http://134.76.20.153/egm/index.php?id=146&L=1&L=1

6 For a deeper understanding of our methodological approach and the setting of our empirical study, see Wöhlke et al. 2019.

References

Andorno, R. 2004. The right not to know: an autonomy-based approach. *Journal of Medical Ethics* 30: 435–40.

Arribas-Ayllon, M., Featherstone, K. and Atkinson, P. 2011. The practical ethics of genetic responsibility: Non-disclosure and the autonomy of affect. *Social Theory & Health* 9:3–23.

Atkinson, P., Featherstone, K. and Gregory, M. 2013. Kinscapes, timescapes and genescapes: families living with genetic risk. *Sociology of Health & Illness*, 35,8: 1227–41. doi: 10.1111/1467-9566.12034.

Bayliss K., Raza K., Simons G., Falahee M., Hanson M., Starling B. and Stack R.J. 2016. Perceptions of predictive testing for those at risk of developing chronic inflammatory disease: a meta-synthesis of qualitative studies. *Journal of Risk Research*, 21,2:167–89. doi: 10.1080/13669877.2015.1119183.

Beier, K., Jordan, I., Wiesemann, C. and Schicktanz, S. 2016. Understanding Collective Agency in Bioethics. *Medicine, Health Care and Philosophy*, 19,3:411–22.

Borry, P., Shabani, M. and Howard, H.C. 2014. Is there a right time to know? The right not to know and genetic testing in children. *J. Law Med. Ethics*, 42:19–27. doi: 10.1111/jlme.12115.

Burke, W. 2014. Genetic tests: clinical validity and clinical utility. *Curr Protoc Hum Genet*, 81:9.15.1–8. doi: 10.1002/0471142905.hg0915s81.

Clarke, A., Richards, M., Kerzin-Storrar, L., Halliday, J., Young, M.A., Simpson, S., Featherstone, K., Keenan, K., Lucassen, A., Morrison, P., Quarrell, O. and Stewart, H. 2005. Genetic professionals' reports of nondisclosure of genetic risk information within families. *European Journal of Humanities Genetics: EJHG*. 13.556-62.10.1038/sj.eigh.5201394.

Crouch, R. and Elliot, C. 1999. Moral agency and the family: The case of live related organ transplantation. *Cambridge Quarterly of Health Care Ethics*, 8:275–87.

Daly, M.B., Montgomery, S., Bingler, R. and Ruth, K. 2016. Communicating genetic test results within the family: Is it lost in translation? A survey of relatives in the randomized six-step study. *Familial Cancer*, 15:697–706. doi: 10.1007/s10689-016-9889-1.

Donchin, A. 2000. Autonomy and Interdependence. Quandaries in Genetic Decision Making. In: Stoljar, N., Mackenzie, C. (eds.), *Relational Autonomy: Feminist perspectives on autonomy, agency and the social self*, New York: Oxford University Press, 236–56.

Etchegary, H. and Fowler, K. 2008. 'They had the right to know.' Genetic risk and perceptions of responsibility. *Psychology & Health*, 23,6:707–27. doi: 10.1080/14768320701235249.

Falahee, M., Finckh, A., Raza, K. and Harrison, M. 2019. Preferences of patients and at-risk individuals for preventive approaches to rheumatoid arthritis. *Clinical Therapeutics*, 41,7:1346–54. https://doi.org/10.1016/j.clinthera.2019.04.015.

Falahee M., Simons G., Raza K. and Stack R.J. 2016a. Healthcare professionals' perceptions of risk in the context of genetic testing for the prediction of chronic disease: A qualitative metasynthesis. *Journal of Risk Research*, 21,2:129–66. doi: 10.1080/13669877.2016.1153503.

Falahee, M., Simons, G., Buckley, C.D., Hansson, M., Stack, R.J. and Raza, K. 2016b. Patients' perceptions of their relatives' risk of developing rheumatoid arthritis and of the potential for risk communication, prediction and modulation. *Arthritis Care & Research*, 69,10:1558–65. doi: 10.1002/acr.23179.

Farmer, T.J. 2010. Relational Obligations: Defending a Non-Voluntarist Argument for Special Responsibilities. *Stance*, 3:39–46.

Featherstone, K., Atkinson, P., Bharadwaj, A. and Clarke, A. 2006. *Risky relations. Family, Kinship and the new genetics*, Oxford: Berg.

Ferrer R. and Klein, W.M. 2015. Risk perceptions and health behavior. *Current Opinion in Psychology*, 1,5:85–9. doi: 10.1016/j.copsyc.2015.03.012.

Gaff, C., Clarke, A., Atkinson, P., Sivell, S., Elwyn, G., Iredale, R. Thornton, H., Dundon, J., Shaw, C., Edwards, A. 2007. Process and outcome in communication of genetic information within families: A systematic review. *European Journal of Human Genetics*, 15,10:999–1011.

Gaff, C.L., Collins, V., Symes, T. and Halliday, J. 2005. Facilitating Family Communication About Predictive Genetic Testing: Probands' Perceptions. *Journal of Genetic Counseling*, 14,2. doi: 10.1007/s10897-005-0412-3.

Gilbar, R. (2007). Patient Autonomy and Relatives' Right to Know Genetic Information. *Med Law*, 26:677–97.

Gilbar, R., Shalev, S., Spiegel, R., Pras, E., Berkenstadt, M., Sagi, M., Ben-Yehuda, A., Mor, P., Perry, S., Zaccai, T.F., Borochowitz, Z. and Barnoy, S. 2016. Patients' attitudes towards disclosure of genetic test results to family members: The impact of patients' sociodemographic background and counseling experience. *Journal of Genetic Counseling*, 25,2: 314–24.

Hallowell, N. 1999: Doing the right thing. Genetic risk and responsibility. *Sociolog of Health & Illness*, 21,5:597–621.

Hallowell, N., Foster C, Eeles R., Ardern-Jones A., Murday V. and Watson M. 2003. Balancing autonomy and responsibility: the ethics of generating and disclosing genetic information. *J Med Ethics*, 29:74–83.

Honneth, A. and Rössler, B. (eds.) 2008. *Von Person zu Person. Zur Moralität persönlicher Beziehungen*. Frankfurt am Main: Surkamp.

Jeske, D. 1998. Families, friends and special obligations. *Canadian Journal of Philosophy*, 28,4:527–56.

Juengst, E.T. 1999. Genetic testing and the moral dynamics of family life. *Public Understan. Sci.*, 8:193–205.

Juth, N. 2014. The Right Not to Know and the Duty to Tell: the Case of Relatives. *Journal of Law, Medicine, and Ethics*, 42,1:38–52.

Kalokairinou, L., Howard, H.C., Slokenberga, S. Kalokairinou L., Howard H.C., Slokenberga S., Fisher E., Flatscher-Thöni M., Hartlev M., van Hellemondt R., Juškevičius J., Kapelenska-Pregowska J., Kováč P., Lovrečić L., Nys H., de Paor A., Phillips A., Prudil L., Rial-Sebbag E., Romeo Casabona C.M., Sándor J., Schuster A., Soini S., Søvig K.H., Stoffel D., Titma T., Trokanas T., Borry P. 2018. Legislation of direct-to-consumer genetic testing in Europe: a fragmented regulatory landscape. *Journal of Community Genetics*, 9,2:117–32.

Kihlbom, U. and Munthe, C. 2017. Health Care Decisions and the Moral Web of Family Responsibilities. In: Lindemann, H., McLaughlin, J., Verkerk, M. (eds.), *Where families and health care meet*, Oxford: Oxford University Press.

Lautenbach, D.M., Christensen, K.D., Sparks, J.A. and Green, R.C. 2013. Communicating genetic risk information for common disorders in the era of genomic medicine. *Annual review of genomics and human genetics*, 14:491–513. doi: 10.1146/annurev-genom-092010-110722.

Leonard, S.J. and Newson, A.J. 2010. Ethical Perspectives. In: Gaff, Bylund (ed.), *Family Communication about Genetics. Theory and Practice*, Oxford University Press, 199–214.

McAllister, M. 2003. Personal theories of inheritance, coping strategies, risk perception and engagement in hereditary non-polyposis colon cancer families offered genetic testing. *Clin. Genet*, 64:179–89.

Mendes, A., Chiquelho, R., Santos, T. and Sousa, L. 2015. Supporting families in genetic counselling services: A multifamily discussion group for at risk colorectal families. *Journal of Family Therapy*, 37,3:343–60.

Mendes, A. Metcalfe, L., Panque, M., Sousa, L., Clarke, A.C. and Sequeiros, J. 2017. Communication of Information about Genetic Risk: Putting Families at the Center. *Family Process*, x:1–11.

Mendes, A., Paneque, M., Sousa, L., Clarke, A. and Sequeiros, J. 2016. How communication of genetic information within the family is addressed in genetic counselling: A systematic review of research evidence. *European Journal of Human Genetics*, 24: 315–25.

Meslin, E.T. and Juengst, E.M. 2019. Sharing with Strangers. Governance Models for genomic Research in a Territorial world. *Kennedy Institute of Ethics Journal*, 29,1:67–95

Mosor E., Stoffer-Marx M., Steiner G., Raza K., Stack R.J., Simons G., Falahee M., Skingle D., Dobrin M., Schett G., Englbrecht M., Smolen J.S., Kjeken I., Hueber A.J. and Stamm, T.A. 2019. I would never take preventive medication! Perspectives and information needs of people who underwent predictive tests for rheumatoid arthritis. *Arthritis Care and Research. Accepted Author Manuscript.* doi: 10.1002/acr.23841.

Rexhepi, H., Åhlfeldt, R.-M., Cajander, Å. and Huvila, I. 2018. Cancer patients' attitudes and experiences of online access to their electronic medical records: A qualitative study. *Health Informatics Journal*, 24,2:115–24. doi: 10.1177/1460458216658778.

Rhodes, R. 1998. Genetic Links, Family Ties, and Social Bonds. Rights and Responsibilities in the Face of Genetic Knowledge. *Journal of Medicine and Philosophy*, 231:10–30.

Ross, L.F., Saal, H.M., David, K.L. and Anderson, R.R. 2013. Technical report: Ethical and policy issues in genetic testing and screening of children. *Genetics in Medicine*, 15,3: 234–45.

Scheffler, S. 1997. Relationships and responsibilities. *Philosophy and Public Affairs*, 26:189–209.

Schicktanz, S. 2018. Genetic risk and responsibility: Reflections on a complex relationship. *Journal of Risk Research*, 21,2:236–58.

Schweikard, D.P. 2015. Quid pro quo? Zur normativen Struktur von Familienbeziehungen. *Zeitschrift für praktische Philosophie*, 2,2:273–310.

Simons G., Belcher J., Morton C., Kumar K., Falahee M., Mallen C.D., Stack R.J. and Raza, K. 2016. Symptom recognition and perceived urgency of help seeking for the symptoms of rheumatoid arthritis, compared with bowel cancer and angina: A mixed method approach. *Arthritis Care and Research*, 69,5:633–1. doi: 10.1002/acr.22979.

Simons G., Mason A., Falahee M., Kumar K., Mallen C.D., Raza, K. and Stack R.J. 2017. Qualitative exploration of illness perceptions of rheumatoid arthritis in the general public. *Musculoskeletal Care*, 15,1:13–22. doi: 10.1002/msc.1135.

Smith, P. 2010. Family responsibility and the nature of obligation. In: Scales, et al. (eds.) *The ethics of the family*, Cambridge: Cambridge Scholars Publishing, 41–58.

Sparks J.A., Iversen, M.D., Yu, Z., Triedman, N.A., Prado, M.G., Miller Kroouze, R., Kalia, S.S., Atkinson, M.L., Mody, E.A., Helfgott, S.M., Todd, D.J., Dellaripa, P.F., Bermas, B.L., Costenbader, K.H., Deane, K.D., Lu, B., Green, R.C. and Karlson, E.W. 2018. Disclosure of personalized rheumatoid arthritis risk using genetics, biomarkers, and lifestyle factor to motivate health behavior improvements: A randomized controlled trial. *Arthritis Care and Research*, 70,6:823–33. doi: 10.1002/acr.23411.

Stack, R., Stoffer, M., Englbrecht, M., Mosor, E., Falahee, M., Simons, G., Smolen, J., Schett, G., Buckley, C., Kumar, K., Hansson, M., Hueber, A., Stamm, T. and Raza, K. 2016. Perceptions of first degree relatives of patients with rheumatoid arthritis about their personal risk of, and predictive testing for, rheumatoid arthritis: A cross-European qualitative study. *BMJ Open*, 6(6):e010555.

Timmermans, S. and Buchbinder, M. 2010. Patients in waiting: Living between sickness and health in the genomics era. *Journal of Health & Social Behavior*, 51,4:408–23.

Townsend, A., Adam, S., Birch, P.H., Lohn, Z., Rousseau, F. and Friedman, J.M. 2012. 'I Want to Know What's in Pandora's Box': Comparing Stakeholder Perspectives on Incidental Findings in Clinical Whole Genome Sequencing. *American Journal of Medical Genetics*, A 158A:2519–25. doi: https://doi.org/10.1002/ajmg.a.35554.

van Boheemen L. and van Schaardenburg, D.J. 2019. Predicting rheumatoid arthritis in at-risk individuals. *Clinical Therapeutics*, 41,7:1286–98.

Verkerk, M.A., Lindemann, H., McLaughlin, J., Scully, J.L., Kihlbom, U., Nelson, J. and Chin, J. 2015. Where families and healthcare meet. *Journal of Medical Ethics*, 41:183–85.

Walter, F.M. and Emery, J. 2005. 'Coming down the line'—patients' understanding of their family history of common chronic disease. *Annals of Family Medicine*, 3,5:405–14.

Wertz, D.C. and Fletcher, J.C. 2004. *Genetics and Ethics in Global Perspectives*, New York: Springer Nature.

Wiesemann, C. 2010. The moral challenge of natality: Towards a post-traditional concept of family and privacy in repro-genetics. *New Genetics and Society*, 29:61–71.

Wöhlke, S. and Perry, J. 2019. Responsibility in dealing with genetic risk information. *Soc Theory Health*. https://doi.org/10.1057/s41285-019-00127-8.

Wöhlke, S., Schaper, M. and Schicktanz, S. 2019. How do moral uncertainty influence lay people's attitudes and risk perceptions concerning predictive genetic testing and risk communication. *Frontiers in Genetics*. https://doi.org/10.3389/fgene.2019.00380.

Wöhlke, S., Schaper, M., Oliveri, S., Cutica, I., Spinella, F., Steinberger, D., Pravettoni, G. and Schicktanz, S. 2020. German and Italian users of web-accessed genetic data: Attitudes on personal utility and personal sharing preferences. Results of a comparative survey ($n = 192$). *Frontiers in Genetics*.

Young, I.M. 1997. *Intersecting Voices. Dilemmas of Gender, Political Philosophy, and Policy*, Princeton, NJ: Princeton University Press.

Part IV

Social and policy implications

8 Direct to consumer personal genomic testing and trust

A comparative focus group study of lay perspectives in Germany, Israel, the Netherlands and the UK

Manuel Schaper, Aviad Raz, Marie Falahee, Karim Raza, Danielle Timmermans, Elisa Garcia Gonzalez, Silke Schicktanz and Sabine Wöhlke

Introduction

In November 2018, genetics start-up Nebula Genomics was reported by 'The Scientific American' to be offering whole genome sequencing to consumers for free in exchange for a case-by-case consent to use the acquired data for research(Weintraub 2018). Such news illustrates how fast-paced progress has been in genetic sequencing technologies and how ubiquitous internet access has driven the development of consumer genomics, allowing companies to offer commercial genomic testing directly to consumers, including testing for rare monogenetic disorders, preconceptional carrier screening, predisposition to common complex disease, and tests providing information relevant to ancestry, nutrition and fitness.[1] According to recent estimates, direct to consumer personal genomic testing (DTC PGT) services like these, especially ancestry tests, have continued to gain popularity, reaching a total of over ten million consumers worldwide (Khan and Mittelman 2018).

DTC PGT has raised a variety of critical questions among experts relating to informed consent, potential negative consequences, clinical or personal utility and the quality and transparency of information provided by DTC PGT companies (Covolo et al. 2015; European Society of Human Genetics 2010). Other analysts have raised concerns about data privacy and unnecessary and/or unsustainable use of health services by consumers after receiving test results (Plothner et al. 2017). Positive impacts of DTC PGT have also been highlighted, such as facilitated access to personal health information (Hogarth, Javitt and Melzer 2008). There has also been an expectation that DTC PGT may improve health-related decision-making and lead to adoption of healthier lifestyles – however, there is currently little evidence to support positive health-related outcomes

(Egglestone et al. 2013; Covolo et al. 2015; Roberts et al. 2017; Bloss, Schork and Topol 2011; Gordon et al. 2012; Carere et al. 2016). Those perspectives on the pros and cons of DTC PGT are the result of studies and reasoning by scholars from medicine, human genetics and medical ethics examining the field with their respective expertise. This study aims at the non-expert perspectives of lay people in regard to those disciplines.

In this study, we examine lay perspectives on DTC PGT in four developed high-income countries: Germany, Israel, the Netherlands and the UK.[2] While the four countries have different regulations relating to genetic diagnostics, consumers are able to obtain genetic test kits from abroad via the internet. Therefore, residents of countries with more restrictive regulation, such as Germany, can obtain test kits via mail order and thus bypass local restrictions (cf. Borry et al. 2012). In this way health-services are transgressing national borders, thus resembling certain forms of health care tourism. 23andMe, for example, while located in the US, ships its test kits to all the countries we discuss in this article, and dozens of others, predominantly in Europe (23andMe 2019). The focus of this article is on differences in attitudes towards and perceptions of DTC PGT by members of the public, i.e. lay people, who have access to clinical genetic testing in their countries, at least in specific medical contexts.

Since we can assume DTC PGT is a hybrid field of commercial endeavour and provision of health care service, trust is an ethical key aspect in relation to its perception by the public. Trust is of critical importance for medical and health care organizations to function, since any person who receives treatment or health care services must to some extent believe in and rely on the trustworthiness of the persons and institutions that provide it (Beauchamp and Childress 2013). This is especially true for medical contexts, in which persons are vulnerable, often in very intimate ways, and need external expertise they can rely on to support health related decision-making, sometimes in critical situations. This creates a duty of care for institutions providing such information (Goold 2001; 2002). While there is debate about the question of whether institutions can be understood as moral actors and whether the concept of trust as an interpersonal relationship can be applied to an organization (Goold 2001), we use here a basic definition of trust based on attitudes towards and expectations of behaviour of another entity:

> Trust is a confident belief in and reliance on the moral character and competence of another person [author's note: or institution] [...]. Trust entails a confidence that another will reliably act with the right motives and feelings and in accordance with appropriate moral norms.
>
> (Beauchamp and Childress 2013, pp. 39–40)

Therefore, we may assume that attitudes towards DTC PGT depend to a great extent on the perception of its providers as being *trustworthy* by the criteria of their (perceived) moral character and competence to provide their services. The perceived *trustworthiness* is then a matter of external judgement by the public as potential or actual consumers of DTC PGT services. In this regard it is interesting to examine how the commercial and clinical contexts of provision of genetic testing compare in terms of trustworthiness in lay people's eyes, or if there are other comparative perspectives for lay people, e.g. other services provided online.

Relevance of lay perspectives

DTC PGT has developed in parallel with a new preventive and predictive paradigm in medicine based on knowledge about human genetic makeup, and its marketing by commercial providers explicitly addresses (healthy) individuals (European Society of Human Genetics 2010). Consumer perspectives on DTC PGT may differ from expert opinion and are likely to be shaped more by cultural and personal context and spontaneous reactions. Discourse on the ethics and regulation of DTC PGT should be informed by the perspectives of all stakeholders, including potential consumers. It is important that the perspectives of members of the public are accounted for in ethical reasoning about commercial genetic testing to counterbalance epistemic imbalance and increase justice in terms of discourse participation (Schicktanz 2012). By choosing a comparative international focus, this study explores diversity of lay perceptions across different countries. We assume that different cultural and social backgrounds have an impact on perceptions and expectations of DTC PGT, and that these differences should be considered in policy making, especially in the context of aspirations to harmonize the regulatory landscape of genetic testing across Europe (Kalokairinou et al. 2018; EuroGentest 2019). Only very few studies of lay perspectives have focused on the public's perception of DTC PGT, and even fewer have captured data from the countries relevant for this chapter. The choice of the four countries discussed in this chapter is based on existing collaborative partnerships between the researchers involved, yet a number of reasons make them interesting examples to study: all four countries we focus on here provide universal health care in a technologically advanced socialized health care system. However, what sets them apart are different cultural backgrounds and some notable differences regarding regulation of (commercial) genetic testing and the structure of the health care system in terms of funding and coverage (Blümel and Busse 2019; Rosen and Waitzberg 2019; Wammes et al. 2019; Thorlby and Arora 2019).

Given the lack of empirical data on lay perspectives of PGT we seek to explore differences in perceptions and expectations of DTC PGT and the ethical implications that come along with it between participants in each of these countries and reflect on their possible relation to the different socio-cultural contexts. Special focus is given to viewpoints on regulation of DTC PGT, its perceived utility, and related data protection and privacy issues and how they relate to and reveal trust and distrust in commercial provision of genetic testing.

Data collection and analysis

For the purpose of this study, 16 focus groups (FG, four for each country) with lay people were conducted (participant distribution: Germany $n=22$, Israel $n=25$, the Netherlands $n=31$, UK $n=21$).[3] The FGs were composed of diverse groups in terms of age, gender and educational background and followed a structured discussion guideline that was initially developed by the Göttingen research group (MS, SW, SS). The FGs were conducted June–November 2016 in Germany, June–October 2018 in Israel, June–July 2018 in the Netherlands and September–October 2017 in the UK.

All FGs followed the same procedure and used the same presentation and information materials, translated by the local research teams. Participants were introduced to the concept of DTC PGT and were shown screenshots of the main pages of different DTC PGT company websites (in their native language when available, or English) to facilitate discussion about general attitudes towards commercial genetic testing. To initiate further discussion, two sample test reports with fictional numbers were shown that consisted of a table indicating individual risk estimates for a range of different conditions (e.g. coronary artery disease, lung cancer, psoriasis, hair loss) in comparison with an average population risk. All FGs were moderated by at least two experienced researchers in each country. For data analysis, we employed a qualitative content analysis approach (Krippendorf 2013) which was facilitated by the software ATLAS.ti. Coding followed a deductive approach, using codes based on common topics present in the medical-ethical literature on DTC PGT, e.g. clinical and personal utility, regulation, privacy concerns.

Results

The analysis focused on a set of pre-identified topics: 1) positive or negative moral and social issues; 2) personal and clinical utility of DTC PGT test results; 3) data protection and privacy; and 4) regulation and governance. All these topics are inherently related to the issue of trust in DTC PGT providers and their services. In the following, we present the results for each country, prioritizing the most prominent viewpoints and then synthesizing the results across countries in a comparative analysis. We mostly refer to attitudes and arguments that have been expressed by the majority (more than 50 per cent of participants) within and across groups in each country or that are unique for a country. Minority positions (significantly less than 50 per cent of participants, small minority = only one or two participants) are indicated when relevant. An overview of the main results is given in Table 8.1.

Germany

German participants expressed a strong distrust towards DTC PGT providers in contrast to physicians. They perceived the latter to be better qualified to manage genetic risk information, provide proper medical consultation and as being obligated to act in the patient's interest:

> [I would] by no means accept an internet offer. There are other experts for that. I trust them more.
>
> GF17L (FG GER2)

> Especially the point about the medical consultation is, I think, enormously important, especially when it comes to something like that, [...] concerning health.
>
> GM17XS (FG GER 4)

It is noteworthy that regardless of the potential benefits of DTC PGT mentioned by some (e.g. bypassing their physician) there was almost a full consensus in the German FGs that the participants themselves would not use health-related DTC PGT services. They expressed a more positive attitude towards lifestyle-related DTC PGT however, especially nutrigenomic and ancestry testing, but stated that personalized disease risk estimates would have a negative effect on personal well-being and be of limited benefit to them. Most participants expressed doubt about the legitimacy of DTC PGT companies and their ability to deliver reliable results:

> Well, I think they make, they only want to make money. I don't know whether they're reputable, and they can write anything. At the hospital, I'd still think there's a laboratory behind it, they really do it correctly. Here it could also be the neighbour who has some chemistry laboratory and then writes down some data for me.
>
> GF19M (FG GER2)

Thus, there is implicit concern that DTC PGT may be fraudulent. Concerns about data protection and privacy were frequently expressed, too. Participants were worried about the future use of their genetic information and concerned that DTC PGT companies might use data for profit, or even to the consumers' disadvantage, as the following quotes illustrate:

> I think we're back at the point companies then, which I mentioned earlier, because the data are then, well, mhm, first of all anonymized again. But in this sense, the data is sold for a lot of money, my data, that is, and I think I wouldn't want that in this sense, either. [...] If I had the choice now to inhibit it that the data are passed on, I'd definitively tick it.
>
> GM17XS (FG GER4)

> I think that should belong into medical hands and not to some private companies, you would also be scared there what else they can do with the data, whether they really are that reputable.
>
> GF25M (FG GER4)

While there is explicit mention of third-party use of the data, there is unspecified concern about 'what else they can do'. Adding to what could be interpreted as an anti-commercial sentiment underlying those quotes, a few German participants distrusted DTC PGT providers because of their online presence:

> Yes. I mean, [...] it doesn't seem very reputable, I must say. [...] I think it's too early for something like that, to shift it onto the internet [...]. It has to develop, as I see it. I mean, I think, if you look at the sites now, then you rather make fun of it. You think that can't be reputable.
>
> GM17XS (FG GER4)

Online provision of genetic testing via a website was seen here as inferior to face-to-face business, while personal contact and consultation was seen as more trustworthy in terms of data protection and accountability of service providers. Regarding its regulation, participants expressed a strong preference for DTC PGT to be legally available yet regulated with appropriate laws to meet consumers' need for reliability and accountability of DTC PGT service providers, especially in the context of risk prediction for serious diseases. As one participant put it:

> I think it's actually even right that you offer something like this so that you have the chance. But as mentioned before, as to cancer, I don't know whether I wouldn't prefer legal framework conditions then.
>
> GF18S (FG GER2)

Here, explicit legal conditions are preferred to prevent possible harms from knowing about personal risk information.

Israel

Unlike the German participants, the Israeli participants were less concerned and generally more positive regarding DTC PGT, and also interested not only in lifestyle-related nutrigenomic and ancestry testing, but in health-related testing as well. The following quotes illustrate this briefly:

> Great, why not. They give out some kind of service, if you want to you can do it, if you don't you don't have to. Sounds good.
>
> IL2M3S (FG IL2)

> But this is also about accessibility, speed and so on. You may be able to do the tests in a public institute, but it may take years. Here it's true you need to pay money, but you get a guarantee of how long it will take and so on.
>
> IL4F6M (FG IL4)

> Everything is about price. Lifestyle I would probably do today, but maybe diseases as well if it is cheap enough. If it's available and cheap, I'll do everything.
>
> IL1M4L (FG IL1)

Interestingly, the positive views are based on perceived aspects of efficiency, accessibility and affordability. Some participants showed appreciation for the possibility of using acquired disease risk information to improve health decisions and behaviour. When confronted with a set of hypothetical disease risk estimates, one participant said:

> I don't know, this test isn't that disturbing or relevant in my eyes. If I'd get higher risks, and it would be about things I can prevent or affect in a way, like heart diseases and so on, I would consider the test helpful.
>
> IL4F5XS (IL FG4)

However, reservations regarding health-related DTC PGT were also expressed. DTC PGT was seen as lacking clinical utility because it did not have added value above and beyond other available forms of diagnostics and risk information. One participant expressed a very critical view, stating that,

> I wouldn't pay for this thing. I think I can create my own chart of risks, more or less. And even then, it's not worth it. I wouldn't do anything about it.
>
> IL1M4L (FG IL1)

Others stressed that the knowledge of personal disease risk estimates would be burdensome and have a negative impact on personal lives, e.g. causing anxiety, doubting the clinical utility of predictive testing or the value of its results as life-changing news:

> My first thought would be that it's all nonsense. Should I start worrying now? It's hard for me to answer this question because I wouldn't do the test in the first place. I believe you shouldn't worry about things prematurely, if they happen, deal with them then. If not, enjoy life, don't spend it worrying about things that might happen. I also have a chance to get run over by a car, I won't start calculating if I should go out or not statistically. You should do what makes you feel good, worry about the consequences later.
>
> IL4F3L (FG IL4)

The most striking result in comparison with FGs from the other countries was the fact that a small minority of participants expressed the view that the commercial aspect of DTC PGT builds and consolidates trust. Some participants stressed that companies need to act responsibly in order to be successful, that the contract's terms and conditions provided legal reassurance:

> I think you are assuming things here without knowing the facts. I don't see a reason for a company to lie to you so bluntly, it's too risky for her as there's a contract and you can sue her. It's true that it's for profit, but I think you are way too untrusting. I'm sure the company will promise anonymity in the contract. They would lose a lot of customers if they won't promise such a thing, so it just makes sense.
>
> IL2M3S (FG IL2)

In this view, the suspected lack of a reliable moral character as a prerequisite for trust is compensated for by the shared knowledge that one can defend oneself against potential fraud – which in turn leads to an expectation of a code of conduct (encompassing privacy) that makes DTC PGT companies successful long-term as trustworthy partners, be it only to avoid negative consequences. Also, participants implicitly expressed trust by talking more about the cost–benefit ratio and regarding DTC PGT as a somewhat regular consumer product. Concerns and expectations regarding DTC PGT revolved more around the acceptability of a service or product in terms of value-for-money, as illustrated by the following quotes:

> I think that if I'm not thinking about the cost of the commercial genetic test, I would definitely order one.
>
> IL2M3S (FG IL2)

> I agree, for the lifestyle one I wouldn't pay money. If I'd receive it for free, why not, but that's it.
>
> IL4F6M (FG IL4)

Participants were concerned about data protection and privacy, especially the confidentiality of data, saying that the data would not be safe with DTC PGT companies and that there would be a risk of discrimination if it was passed on to third parties such as employers, thus expressing distrust in the business practices. A few participants also stressed that there was some extra value of the data for the companies in case it is used for research, insinuating a motive to do more with the data than using it to provide the advertised services. The following quotes illustrate those concerns:

> [...] I don't believe the information will be safe, so I wouldn't do it personally. Maybe if they can prove that they only have information about my lifestyle, and then I might not mind if it's stolen. But I think they can get a lot more than that with my spit sample, so it's not safe.
>
> IL1F3S (FG IL1)

> [...] it might affect your life in unexpected ways. Like employers not hiring you for having a high risk for cancer. Or even affect your military service.
>
> IL1M1XS (FG IL1)

Another participant expressed that DTC PGT companies might even use their data to create incentives to purchase additional services to increase sales.

> [...] they can lure people to spend more money. If for example you wanted a lifestyle test, and then they tell you they found something unrelated to lifestyle but might still be important, but you must pay more to get it. They can potentially scare customers into spending more money, and I think that's unethical. They can also group up with other drug companies and write on my test results that they recommend that I start taking this and that or use that shampoo.
>
> IL2F6XS (FG IL2)

Here, a distrust is expressed based on the potential for excessive exploitation of consumers who have already purchased a genetic test. Israeli participants expressed varying views regarding regulation of DTC PGT. It was common to argue for it to be legal, though most felt that some sort of regulation was necessary, e.g. to prevent discrimination on the labour market or abusive business practices. This nuanced pattern of preferences regarding regulation is best exemplified by the following quote:

> I think it should be legal, but I also think we need to regulate and supervise it. We need to check the company; the people working there and make sure they are doing it correctly and morally. Need to make sure they have all of the qualifications. We need to make sure the information only gets to where it's supposed to. As a state we need to make a law that regarding genetic tests they are only allowed to bring the information to the customer and no one else. That way we might be able to stop all the bad scenarios we suggested earlier.
>
> IL2F5M (FG IL2)

Interestingly, in this example and the previous one, the business conduct is discussed using the terms 'unethical' and 'morally', while the latter is also mentioning the necessity of proper qualifications as condition for trust. Some Israeli participants even advocated against regulation of the DTC PGT market and providers, which is a unique finding for the four countries. These participants were being optimistic about the market's development or saying that regulation in their country would be desirable in theory but ineffective in reality, as one participant put it:

> Unfortunately, I'm not a big fan of regulation in this country. It's a very good idea in general, but in Israel it doesn't work properly. If I was German my answer would probably have been different. So, regulation is fantastic, but it doesn't work here, so I'm against regulation in this case. And it should be legal. Let them profit, if they succeed.
>
> IL4M4L (FG IL4)

> What are we, a dictatorship? We will forbid these tests? No, if someone wants to do it, he should be able to. They should be legal.
>
> IL3F6XS (FG IL3)

The views of the Israeli participants thus indicate a higher readiness to accept health-related DTC PGT as a consumer good among others, while potential problems or implications with it are not perceived as a good reason to restrict its possibilities legally.

The Netherlands

Dutch participants expressed strong reservations regarding health-related DTC PGT. Their criticism focused on a perceived lack of diagnostic quality (i.e. analytical validity, even though participants did not use such terms) provided outside of the regular health care system and the possible burden of risk information. The following snippet from one of the discussions illustrates this very well:

> Actually, it has no use. You can better do a qualified test right away.
>
> NLM18M (FG Nl3)

> [In response:] What I think about tests like that, I mean look, we were just talking about people who have a genetic disease in the family and they want to see if they have it too. You can imagine something like that. But just, these are my genes, find out what I could have and the chance that... That is so useless! Because we all have something. We are all probably going to die sometime.
>
> NLM21L (FG NL3)

While not using the term, participants referred here to a lack of clinical utility. Participants strongly preferred health-related PGT services to be provided by physicians. They doubted that DTC PGT companies were appropriately qualified to reliably produce and deliver genetic risk information to consumers, as several participants pointed out:

> You get a chance of 71 per cent at agency A, and then I go to agency B and I do not know if I get the same result. I do not know to what extent it is science. Then you can decide: I take A because the outcome is smaller!
>
> NLF26M (FG NL4)

> I would take it to the clinical geneticist anyways.
>
> NLF10L (FG NL2)

So, while they did not expect DTC PGT companies to deliver reliable information, or different companies to deliver the same information for the same case, it became also apparent that they need external expertise.

Participants expressed an interest in ancestry testing, suggesting that the quality of the service was far less important for this service. They further doubted that there is much health benefit in receiving this information and expected that it would be burdensome to an individual (e.g. anxiety, psychosocial stress). Furthermore, the Dutch participants were the only ones who emphasized the potential danger of interpreting risk information that indicates low genetic risk as an excuse for an unhealthy lifestyle, as the following discussion bits show:

> You will get false hope.
>
> NLF20L (FG NL3)

> [In response:] Yeah, you will get a lot of false hope.
>
> NLM21L (FG NL3)

> [In response:] A sense of fake security or something like that. You think you are healthy.
>
> NLF17M (FG NL3)

> What I do not like, is that if you look at these results, you can think: 'Oh, I have less than the average risk of lung cancer. Well, then I light ten (cigarettes) up', so to speak. I mean that when people see that their risk is lower, that they will just adopt a worse lifestyle. You only consider your DNA risk and not the environmental factors. I find this scary.
>
> NLF3XS (FG NL3)

Participants were also concerned about data protection and privacy. They discussed a variety of negative hypothetical scenarios including possible discrimination by employers, the possibility of their DNA entering data banks for criminal investigations or companies using their data to pursue other commercial interests. The following statements provide examples for that:

> Could this also go to companies that, for example, have secondary interests? For example, if you are talking about obesity – I actually do not know it very well – or for example the tobacco industry, let's say. Perhaps there could be very strange side-tracks created with all the data. And I do not expect that from the medical sector and certainly not from university research hospitals etc. But here there is a kind of uncontrolled sprawl. And I do not know who would have access to get your data. I would find it very scary. It's nothing for me anyway.
>
> NLF30L (FG NL4)

> Who will have access to the data? Who owns the data? What do you test? Is it reliable?
>
> NLF11XS (FG NL2)

> Exactly! Is it reliable? And what I find very scary is you have to send your DNA and we know DNA is also used for criminal purposes, at least there is a big data bank to investigate crime, imagine that something happens to your DNA! I do not find it verifiable. I find it scary.
>
> NLF30L (FG NL4)

While the imagined case of using genetic information in forensics seems to imply here the idea that submitted DNA could be brought to a crime scene, all of the above statements show a distrust in DTC PGT providers based on a lack of knowledge about their practices, i.e. being unable to judge their moral character and form respective expectations.[4] Given such negative outlooks on the topic in the FG discussions, it is not surprising that a small number of participants even stated that DTC PGT should be prohibited, like the following participant:

> I don't think this should be allowed, because I mean you will go to the doctor: 'look, doctor, I got this results, what are we going to do now?
>
> NLF19M (FG NL3)

Arguments against prohibition were brought forward as well, based on the view that it would infringe on freedom and therefore would require substantial reasons, a view that was also expressed in the Israeli FGs. One Dutch participant said:

> You know, you can't prohibit it, because it is not completely forbidden to test someone's genes. You just have to ask yourself if you should make a big deal out of it, if the result can mean anything. Because it doesn't mean anything. You get information about this disease and that disease, and?
>
> NLM21L (FG NL3)

This view shifts responsible handling of personal genetic information back into the consumer's hands, who must make an informed decision for themselves.

United Kingdom

UK participants were also mostly opposed to health-related DTC PGT, mainly because they perceived a lack of necessary counselling and follow-up consultations. Some expressed the view that health-related genetic testing should not be allowed to be commercialized.

> I personally feel it shouldn't be available online these tests, because you're dealing with people's health, and I know a lot of organizations is becoming big business, but if you're dealing with cases where you know, health is the main element, why would you consider that to be a business of some kind?
>
> UKM4S (FG UK1)

This statement purports the notion that there should in principle be no commercial exploitation of personal genetic information whatsoever. In the UK FGs, there was a lack of trust in DTC PGT providers based on a suspected lack of competence and dedication to the consumer's benefits because of their commercial interests (i.e. a perceived lack of moral character). Participants preferred health-related genetic testing to remain in the realm of the National Health Service (NHS), which was seen as trustworthy because of its dedication to patients'/consumers' interests. The following quotes illustrate how trust in health care institutions and the NHS was repeatedly mentioned:

> I would honestly be expecting to go to the hospital to be getting that test, I wouldn't trust anything I got online unless it clearly stated it was NHS, maybe it was a direct partner with the NHS and they've approved it, I wouldn't trust anything other than that [...].
>
> UKF7S (FG UK1)

> I think if that was offered through the NHS I would trust it a lot more, I would trust that they would keep the data confidential because they have more rules.
>
> UKF15S (FG UK4)

> It would be quite nice to say 'oh well actually I'm fine'. I think for me I definitely would like a test from somebody that I considered, I had confidence in and felt that they had authority you know, and they can place some sort of perspective for me [...].
>
> UKM17L (FG UK3)

A common central concern was related to the confidentiality of genetic data and the possibility that DTC PGT companies might profit from using it for research:

> Pretty much if I was 100 per cent sure that the company that's doing it is a bona fide company, the first thing I would require from them is confidentiality, what confidentiality can they offer me in terms of using my data or my sample for future research, that my identity was not disclosed, and I think at this stage in time there is no guarantee of that, so I would be very, very hesitant to participate in any kind of tests like that.
>
> UKM1L (FG UK1)

Underlying all these is a principal doubt that confidentiality and data security can be reliably guaranteed, while the ability to do so is reported as the main requirement for a trustworthy company. For those reasons, a strong preference for the regulation of DTC PGT business was expressed, while only one participant preferred DTC PGT companies to be banned altogether:

> I think it should be a tool, one tool, but hopefully through some authority you know.
>
> UKM17L (FG UK3)

> I definitely think it needs to be regulated.
>
> UKM5XL (FG UK2)

> Those companies should be buried at birth.
>
> UKM24XL (FG UK4)

Some participants described potential benefits of DTC PGT for certain individuals. One group discussed potential value for persons who are carriers of a genetic disease or who have no possibility of investigating their family history of health problems or diseases. The following quote illustrates this point:

> It might be helpful for those that don't know their families, like if they've been adopted or fostered or whatever, they haven't got the people there to ask.
>
> UKM18XS (FG UK3)

> [In response:] They haven't got the family history, yeah.
>
> UKF11XS (FG UK3)

Without explicitly referring to the terms of clinical or personal utility, there is appreciation for the potential benefits for other people, thus extending the scope beyond own personal benefits. It was also mentioned that DTC PGT may be useful for improving health decisions (e.g. choosing individual health plans), yet participants felt that they would need assistance and trusted external expertise to understand test results.

> I think we'll all end up with our own health plans, I think it's going to move more to that and it will be a do it yourself plan.
>
> UKM5XL (FG UK2)

> That would totally confuse me if that came through my door.
>
> UKF3L (FG UK2)

> [In response:] Yeah.
>
> UKF9S (FG UK2)

Nonetheless UK participants had an interest in lifestyle-related DTC PGT, especially for nutrigenomic and ancestry testing, which was seen as having less potential for negative consequences for users, e.g. when there is a confidentiality breach.

Table 8.1 Side-by-Side Comparison of Main Results

Topic	*Germany*	*Israel*	*Netherlands*	*UK*
Positive or negative moral and social issues	Critical attitude towards health-related PGT – Distrust in PGT providers' quality and intentions – Anti-commercial sentiment regarding genetic testing services	Mixed attitude towards health-related PGT – Mixed attitude in regards to trust in PGT providers – Trust based on market laws – Positive view on PGT based on its efficiency as a provision model	Critical attitude towards health-related PGT – Distrust in PGT providers' quality	Critical attitude towards health-related PGT – Distrust in PGT providers' quality and intentions – Anti-commercial sentiment regarding genetic testing services – Perceived lack of counselling and follow-up consultations
	Strong preference of physicians over commercial PGT providers		Strong preference of physicians over commercial PGT providers	Strong preference of NHS over commercial PGT providers
	Interest in lifestyle PGT, mentions of – Nutrigenomics, ancestry	Interest in lifestyle PGT, mentions of – Nutrigenomics, ancestry, other traits	Interest in lifestyle PGT, mentions of – Ancestry	Interest in lifestyle PGT, mentions of – Nutrigenomics, ancestry
Personal and clinical utility of DTC PGT test results	Pro: – Bypassing physicians (minority position)	Pro: – Improving health decisions		Pro: – Improving health decisions– Potential benefits for certain minorities
	Con: – Personal risk information burdening – Personal risk information obsolete since other forms of diagnostics are available	Con: – Personal risk information burdening – Personal risk information obsolete since other forms of diagnostics are available	Con: – Personal risk information burdening – Personal risk information obsolete since other forms of diagnostics are available – Perception of potential danger of taking risk information that indicates low genetic risk as excuse	Con: – Tests results are confusing, and expertise is required

Data protection / privacy	Concerns about misuse of data and confidentiality Distrust in internet-based marketing of genetic testing services	Concerns about misuse of data and confidentiality Perceived risk of discrimination based on genetic risk information Perceived risk of abusive business practices	Concerns about misuse of data and confidentiality Perception of PGT companies having an unfair extra benefit	Concerns about misuse of data and confidentiality Perception of PGT companies having an unfair extra benefit
Regulation and governance	Preference for regulation of PGT but also its availability for consumers Regulation preferred especially for predictive tests for serious disease	Mixed preferences regarding regulation with a pro-regulation tendency Explicit preference of availability of PGT based on liberal convictions and market optimism	The Dutch groups missed an extensive discussion of preferences regarding regulations due to time restrictions during the focus groups	Preference for regulation of PGT but also its availability for consumers
	Minority position arguing for a ban	Minority position arguing for a ban	Minority position arguing for a ban	Minority position arguing for a ban

Cross-country comparison of main themes

1) Positive and negative social and moral issues: It was apparent that most participants were unfamiliar with DTC PGT. We found a variety of positive views between countries and agreement regarding negative viewpoints. While the German, UK and Dutch participants had a very critical view on health-related DTC PGT, the Israelis were more open towards DTC PGT and had a more liberal view on its regulation. German, UK and Dutch participants tended to prefer genetic testing to be done within a national health system or by a trusted physician, and not by a company guided by market principles. Yet, in the UK the focus was on the NHS as a trustworthy system, while the German and Dutch participants focused on the physician as a trustworthy individual with proper expertise. Strikingly, only in the Israeli FGs did a small minority express that the application of market principles builds and consolidates trust in DTC PGT, while in Germany and the UK the opposite was the case.

2) Personal and clinical utility of DTC PGT test results: Attitudes towards and perceptions of different notions of utility were mixed, while participants made a distinction between disease-related testing and 'lifestyle' tests. Here, Israeli participants were more optimistic about the potential of disease risk estimates to positively affect health behaviours and decision-making than their UK, German and Dutch counterparts, especially regarding the possibility to compensate for risk or prepare for negative developments in case of serious disease such as cancer. However, some UK participants stated that health-related DTC PGT may be a useful service for carriers of genetic disease or persons who have no or only limited access to their family history. Negative impact of predictive genetic information on health because of false interpretation of results, i.e. taking a lower genetic risk as an excuse for a riskier lifestyle, was a concern in the Netherlands. In each country, at least some participants expressed an interest and trust in lifestyle DTC PGT (e.g. relating to ancestry or nutrigenomic testing), which was perceived to have some personal utility and less potential for negative consequences.

3) Data protection/privacy: Concern about the security of consumers' genetic data was common among all countries. However, differences occurred between the countries regarding potential negative consequences. While both UK and Dutch participants speculated that DTC PGT companies might find additional ways to profit from consumer data (e.g. by using it for research or patent development), the German participants were mostly sceptical about existing data protection measures because they perceived online businesses as unsafe and felt that the policies were not transparent. Israeli participants, however, were specifically concerned about genetic discrimination or abusive business practices (e.g. use of genetic information to create pressure on consumers to buy extra services or products).

4) Regulation and governance: It was almost universally preferred in all FGs that DTC PGT is subject to regulations to ensure its safety and reliability (e.g. a system of laboratory certification, standards and/or oversight in order to ensure quality, confidentiality and accountability, and/or restrictions on the kinds of tests performed), but not prohibited. However, only in Israel did some participants express a preference for the unregulated provision of DTC PGT.

Discussion

The expert views on DTC PGT and its ethical implications as outlined briefly in the introduction of this chapter are primarily grounded on reasoning based on professional experience and conclusions drawn from scientific research, balanced out in scientific reasoning and careful argumentation. Such views differ from opinions expressed by lay people, which are more shaped by personal experience and cultural and social background. Such aspects are not reflected in a FG setting, where opinions are uttered spontaneously and in interaction with others. Our results show that there is considerable variation in attitudes towards health-related DTC PGT across the four countries in regard to trust and to the reported or implicit aspects that influence trust in DTC PGT companies and their services. To some extent, this can be explained by participants' situatedness in different health care systems or socio-cultural contexts. There are striking similarities regarding the concerns, especially for privacy, data protection and regulation. These seem to be related to the sensitivity of genetic risk information and its possible negative implications in general. In the following we will discuss the major similarities and differences across the four countries highlighting their relation to trust.

Major similarities

Participants from all countries expressed a sense of *caveat emptor* (Latin for 'let the buyer beware' (Encyclopaedia Britannica 2019) in regard to DTC PGT services that is often based on a careful consideration of the sensitivity of personal data. Interestingly, this finding is consistent with the results of a recent study of Australians' views on the same topic (Metcalfe et al. 2018). The main reasons for caution were related to the perception that online business is generally less trustworthy (in terms of privacy) and that genetic data is particularly sensitive information which requires a high level of trust in a company that is supposed to handle it.

The finding of a widely expressed wish for regulatory oversight of DTC PGT companies across the countries is also consistent with the findings of a US study that examined DTC PGT consumer perspectives (Bollinger, Green and Kaufman 2013). It illustrates an awareness of the potentially problematic aspects of genetic risk information and its possible negative implications for the individual, and that there need to be certain restrictions or legal warranties to DTC PGT business practices that would affect trust in a positive way. However, a consumerist standpoint was evident in that a preference for prohibition was an exception in all FGs. Participants felt that, regardless of how hesitant they personally were, a ban would (and it would in fact) place undue restriction on entrepreneurship and patient/consumer choices (cf. Schaper, Wöhlke and Schicktanz 2018; di Norcia 2012). Across all countries, an interest in lifestyle tests was prevalent. The overall less critical stance regarding lifestyle testing is somewhat surprising, since the main concern in relation to DTC PGT was about data privacy, but the data and samples for lifestyle testing are processed and stored just as they are in health-related testing. This may be because they are perceived as having no explicit connection to disease, so they are framed as less open to potential discrimination. Yet this demonstrates a lack of awareness about the nature and potential implications of genomic data that should be addressed in related consent processes. According to Goold (2001; 2002) trust can be raised by certain factors that do not in fact justify that raise. So, while 'lifestyle' as a label appears to affect trust in a positive way, it is doubtful whether increased trust is actually justified, given that relevant circumstances regarding the handling of genetic information remain the same. Balanced information about both benefits and risks associated with DTC PGT is essential to facilitate informed decision-making by consumers and even the ground for well-placed trust.

A major selling point in DTC PGT companies' advertising is the idea that test results may help to improve health behaviour (Covolo et al. 2015; Einsiedel and Geransar 2009; Goldsmith et al. 2012; Schaper and Schicktanz 2018), and various studies have shown that this is among the main motives of potential and actual PGT consumers (Goldsmith et al. 2012; Henneman et al. 2013; Bayliss et al. 2016; Mavroidopoulou, Xera and Mollaki 2015; Oliveri et al. 2016; Gollust et al. 2012). Our results show that some members of the public in the countries of this study are aware of this potential for benefit. Yet most appear to be distrustful and remain sceptical as to whether they would choose to undergo DTC PGT, whether out of dispositions unrelated to genomic testing itself (e.g. anti-commercial sentiment, general distrust in online business, fear of confidentiality breaches or discrimination) or perceptions related directly to DTC PGT (perceived lack of clinical and personal utility, possible negative impact of test results on personal lives). Further understanding of psychosocial predictors of positive and negative viewpoints is needed.

According to our analysis, there is a near-universal preference for the DTC PGT market to be regulated to secure quality standards and consumer and privacy protection. This viewpoint is probably driven by a wish for freedom of choice for consumers and patients alike, yet it also shows that there is a wish for safe conduct guaranteed by a superior trusted third party, as a guarantor that such businesses can be trusted. This would mean that the dedication to patients'/consumers' benefit as a prerequisite for well-placed trust is substituted. Such views correspond to aspirations to 'harmonize genetic testing across Europe' as employed by the network 'EuroGentest' funded by the European Commission (EuroGentest 2019). The discrepancies in lay perspectives that we observed, may however inform reasoning about the development of regulation at national levels. Our results indicate that preventing access to DTC PGT is perceived as unduly restrictive and a regulated market is preferred that provides measures to protect consumers from fraud, genetic discrimination or psychosocial harm, without restricting access to and provision of such services completely.

Major differences

A notable difference can be observed in participants' perceptions of what could be called *markers of trustworthiness* – perceived aspects that influence the will or ability to trust. As Goold (2001) points out, there are certain key components that influence and/or justify trust, especially knowledge about the trustee's behaviour with a certain degree of predictability and a sense of shared goals and values, similar to what Beauchamp and Childress (2013) call 'acting with the right motives and feelings and in accordance with appropriate moral norms'. Both German, Dutch and UK participants displayed distrust based on these factors, as became evident in their repeated comparison with genetic testing in the clinical context. However, there are interesting nuances: the scepticism of German participants in particular was based more on a perception of online provision of DTC PGT as being untrustworthy compared to provision by physicians, which implies trust on a personal level, whereas the UK participants preferred provision from within or approved by the NHS, which remains an institutional level of trust. Medical professionalism as a property of the physician on the one hand, and the NHS-label as an institutional aspect on the other, served as the main markers of trustworthiness and legitimacy, influencing the likelihood that an individual would consider DTC PGT. This is in line with some older findings about public trust in health care in these countries that characterized trust in England and Wales as very high in comparison with Germany and the Netherlands (van der Schee et al. 2007). The NHS, like other public institutions, relies on public trust to function in its role (Sterckx and Cockbain 2014). For lay people, such reliable, non-commercial institutions of health care provision are familiar and reliable, meeting Goold's (2001) trust criterion of knowledge, focused solely on patient benefit, thus also meeting the criterion of shared values. Since there is easy access to high quality health care in these countries, consumers are less likely to pay for additional commercial services, unless perhaps they are convinced of its clinical or personal utility, safety and reliability.

While there are specific screening programs for some of the most common cancers in the Netherlands, general health checks are otherwise not recommended because there is doubt as to whether there is sufficient benefit (Rijksinstituut voor Volksgezondheid en Milieu 2019; Healthcare for Internationals 2018). Preventive full body examinations via MRI (*total body scan*) are even prohibited (Rijksoverheid 2019; Check de Check 2019), which is why it is common to undergo them in Germany or pay for them. The scepticism we observed in the Dutch FG participants may then be based more on the genetic testing aspect as such, e.g. its perceived limitations, and has less to do with the perception of DTC PGT providers as trustworthy or not. Israeli participants' partial and sometimes explicit expression of trust in DTC PGT providers still stands in contrast to this notion, insofar as they accepted them as a regular option. Here we may assume that a higher openness to DTC PGT is the by-product of a high confidence in the own ability 'to trust wisely' (Goold 2001) when it comes to health-related commercial services, as becomes clearer when we compare the results regarding privacy and data protection: German, but also Dutch and UK, participants expressed greater concerns about the security of genetic data relating to DTC PGT than their Israeli counterparts. Data protection and privacy are highly valued in Germany, which introduced the first data protection laws in history during the 1970s (Federal Data Protection Law) and whose Federal Constitution Court defined *informational self-determination* as a fundamental right of all citizens in a rule about a census in the 1980s (Garstka 2003). The discrepancy with Israel here is consistent with its reputation as an innovation-friendly country with an extremely high number of technology start-ups per capita (Kon et al. 2014).[5] Out-of-pocket payments for medical services like dental care or pharmaceuticals are common in Israel and the country is on the vanguard of implementation of eHealth technologies and services (Rosen and Waitzberg 2019; Weiner 2012). These factors may account for the comparatively positive views on health-related DTC PGT services that imply a consumer perspective. Furthermore, the acceptance of applications of genetic testing in general is relatively high in Israel as well (Raz and Schicktanz 2016; Rosner, Rosner and Orr-Urtreger 2009). The market optimism expressed by some Israeli participants and their strong advocacy for legal availability of DTC PGT corresponds with these findings and connects to another social condition of trust pointed out by Goold (2001), the role of *deterrence*: the notion of a contractual relationship provides reassurance that DTC PGT companies have legal restrictions limiting their activities as well as duties to their customers, who have rights as consumers. In comparison, German, Dutch and UK participants were less likely to take the perspective of potential consumers of medical goods, thus distrusting the service.

German participants, and to a lesser extent the Dutch participants, were very critical regarding health-related PGT and the trustworthiness of DTC PGT companies. However, according to a 2013 study, there is public awareness in the Netherlands of the rise of genetic/genomic testing and positive beliefs regarding its (non-commercial) application (Henneman et al. 2013).

In Germany, genetic testing may be carried out exclusively by medical professionals when there is a specific indication, and out-of-pocket payments play a minor role in health care.[6] Therefore, DTC PGT might be assessed more from the perspective of potential patients who benefit from high standards in medicine in terms of quality of care and a trustful physician–patient relationship with trust-building face-to-face communication – privileges they feel are not given in commercial online provision of genetic tests, where communication works one way and where lay people do not know what to expect and whether providers are committed to their benefit or only their own interests (Schaper, Wöhlke and Schicktanz 2018; Goold 2001). Critical views among the Dutch participants may be rooted in similar experiences of sufficient availability of genetic testing in a controlled health care setting that is restrictive on screening offers, and an overall high trust in health care (van der Schee et al. 2007; Kalokairinou et al. 2018). Their low acceptance might be related to the comparatively high educational level of the participants, who hold more critical views regarding genetic testing altogether (Stewart 2018). Goold (2001) has stated that trust in institutions depends on a certain degree of knowledge, yet too much knowledge may also affect trust negatively. Therefore, a higher educational level may correspond to distrust in case of our Dutch participants. However, German and UK participants expressed an anti-commercial sentiment that might be explained by their personal experience and support of their state's publicly funded health care system that places a high value on equality and solidarity (Evans and Wellings 2017; Robertson, Appleby and Evans 2018; *Deutsches Ärzteblatt* 2017). Thus, the German groups expressed the least appreciation for health-related DTC PGT as a new consumer option, even though they were not in favour of its prohibition, either. Overall, in comparison with the other countries, the German participants were most critical of DTC PGT, and in Israel they were most accepting of services of this kind.

The analysis of trust-related aspects provides important insights into the public's expectations and concerns regarding commercial provision of genetic testing, and we have presented here some non-representative data on the topic. International or at least European regulation of DTC PGT should consider, among other factors, common concerns and protect consumers from possible negative aspects as outlined above and in the literature, especially if further evidence supporting our results occurs in other studies in the future. We have found that health-related genetic information is preferred to be obtained via trusted authorities like health care professionals, health care systems or with health insurance companies as guarantors. Such institutions and their services are trusted more, since in general they already have a record of trustworthy behaviour and the public knows what to expect from them, i.e. can judge their 'moral character' (Beauchamp and Childress 2013). Here, preferences vary across different socio-cultural contexts in regard to what exactly affects and/or justifies trust, and there are indeed different models of provision of genetic testing in the health care setting in place for every country. In accordance with existing regulation and recommendations by professional bodies, guidelines specifically for DTC PGT should be developed with stakeholder involvement, including the industry and the public. While regulation is desired, an explicit or de facto prohibition of DTC PGT per se appears to be unpopular among the public and would need good justification. Harmonization efforts targeted specifically at the marketing and provision of DTC PGT services may be more suitable, e.g. imposing strict rules on its advertising, specifying consumer rights or introducing mandatory involvement of physicians for health-related testing and warnings about possible risks. Such measures may then provide the necessary conditions for public trust in DTC PGT companies to build. As for the perception of them as being trustworthy or not, it is possible that the preliminary state of them and their practices still being unknown to the majority of the public works as a disadvantage, since consumers do not know what to expect. If we transfer the central idea of trust as a prerequisite for medical and health care institutions to function to DTC PGT we may assume that for such enterprises to be successful in the long run it is crucial that they invest in building and maintaining trust as much as medical and health care institutions do. This may then increase the perception of their services as a valid alternative to clinical genetic testing.

Concerns around the security and privacy of genetic information provided by commercial companies are a widespread trust-related issue and consumer expectations may not be aligned with the data protection policies of DTC PGT companies (Christofides and O'Doherty 2016). Participants in all countries seemed unaware or unconcerned that such risks may be also associated with lifestyle testing, which shows the importance of clear and transparent privacy and data protection policies (Hall et al. 2017; Laestadius, Rich and Auer 2017). Yet it also shows that there is a significant difference between influencing trust and establishing conditions that justify it (Goold 2001). This illustrates the importance of appropriate provision of information about the risks as well as the personal and clinical utility of DTC PGT to inform consumers' decision-making. Such information might include a risks and side-effects notification in advertising and on test kits with a recommendation to seek experts for in-depth information, as is common for pharmaceuticals. Such information should be tailored to suit not only the literacy needs and preferences of individuals but also their personal and cultural perspectives.

As this study is qualitative in nature, we cannot claim the results are representative of each country's general population or make definite claims about differences between the countries. Despite efforts to harmonize recruitment procedures, there was some variation across the four countries, for example the use of local research participant recruitment platforms, possibly reaching different audiences. The discussion topic might have attracted more participants who had previous experience with genetic testing than average and who were eager to discuss their experiences, while persons with no previous experience or related knowledge might have been less motivated to participate in a discussion. In a similar vein, participants in all FGs had a high educational level, and perceptions and opinions may not be representative. The age distribution of participants does not reflect the distribution of age groups in each country. However, four FGs were conducted in each country to achieve a balanced mix of viewpoints.

Concluding remarks

We have presented here some evidence from a qualitative study of lay perspectives to fill a gap in the literature and allow for broader and more in-depth discussion of DTC PGT, its risks, benefits and appropriate regulatory approaches. The cross-country comparison is a useful approach to help identify differences and common ground in attitudes and perceptions across the countries included and work out some rather unique details that a quantitative study might not have been able to cover in the same way. The rich and extensive material generated by our focus group study provides the possibility for future in-depth analysis regarding all aspects touched upon here only briefly. For example, it would be interesting to show with more detail how regulation of genetic testing and different health care systems relate to the findings. From a medical ethics point of view, it would be interesting to analyse more deeply the perception of DTC PGT providers' responsibilities in subsequent publications and whether there are differences between the four countries.

Notes

1 According to our last search in early 2017, about 90 companies worldwide were marketing genetic/genomic testing or interpretation of test results to consumers.
2 We discuss the countries in alphabetical order. While the focus is on Europe, Israel is included as a reference point.
3 The study has been approved by the University of Göttingen Human Research Review Committee (Ref. 16/10/14), the Science, Technology, Engineering and Mathematics Ethical Review Committee of the University of Birmingham (Ref. ERN 17-0298) and the Medical Ethical Review Commission of the University Medical Center of the University of Amsterdam (VUmc_2018.084).
4 It is likely that such statements were influenced by media reports of a large-scale mass DNA screening to solve a murder case happening at the time the FGs were conducted (Brown 2018).
5 This is also supported by the fact that Israelis use social media more than any other people in the world (The Times of Israel 2018).
6 According to official statistics, total health care expenditure in Germany exceeded one billion Euro *per day* in 2017, however the Germans spend only one billion Euro *per year* on out-of-pocket health care services (Statistisches Bundesamt 2019; Der Spiegel 2018).

References

23andMe (2019). 'What Countries Do You Ship To?' [Online]. Available at: https://customercare.23andme.com/hc/en-us/articles/360000145307-What-Countries-Do-You-Ship-To- (Accessed 17 April 2019).

Bayliss, K., Raza, K., Simons, G., Falahee, M., Hansson, M., Starling, B. and Stack, R. (2016). Perceptions of Predictive Testing for those at Risk of Developing a Chronic Inflammatory Disease: a Meta-Synthesis of Qualitative Studies. *Journal of Risk Research,* 21(2), pp. 167–89.

Beauchamp, T.L. and Childress, J.F. (2013). *Principles of Biomedical Ethics.* 7th edn. New York: Oxford University Press.

Bloss, C.S., Schork, N.J. and Topol, E.J. (2011). Effect of Direct-to-Consumer Genomewide Profiling to assess Disease Risk. *New England Journal of Medicine,* 364(6), pp. 524–34. doi: 10.1056/NEJMoa1011893.

Blümel, M. and Busse, R. (2019). The German Health Care System. *International Profiles of Health Care Systems*, edited by E. Mossialos, A. Djordjevic, R. Osborn, and D. Sarnak, pp. 69–76. Available at: https://international.commonwealthfund.org/countries/germany/

Bollinger, J.M., Green, R.C. and Kaufman, D. (2013). Attitudes about Regulation among Direct-to-Consumer Genetic Testing Customers. *Genetic Testing and Molecular Biomarkers,* 17(5), pp. 424–28.

Borry, P., van Hellemondt, R.C., Sprumont, D., Jales, C.F., Rial-Sebbag, E., Spranger, T.M., Curren, L., Kaye, J., Nys, H. and Howard, H.C. (2012). Legislation on Direct-to-Consumer Genetic Testing in seven European Countries. *European Journal of Human Genetics,* 20(7), pp. 715–21.

Brown, K.V. (2018). Dutch Police Are DNA Testing 21,500 Men to Solve a 20-Year-Old Murder [Online]. Available at: https://gizmodo.com/dutch-police-are-dna-testing-21-500-men-to-solve-a-20-y-1823562697 (Accessed 8 April 2020).

Carere, D.A., Kraft, P., Kaphingst, K.A., Roberts, J.S. and Green, R.C. (2016). Consumers report lower Confidence in their Genetics Knowledge following Direct-to-Consumer Personal Genomic Testing. *Genetics in Medicine,* 18(1), pp. 65–72.

Check de Check (2019). Total bodyscan [Online]. Available at: https://checkdecheck.nl/wat-is-een-preventieve-zelftest/total-body-scan/ (Accessed 25 April 2019).

Christofides, E. and O'Doherty, K. (2016). Company Disclosure and Consumer Perceptions of the Privacy Implications of Direct-to-Consumer Genetic Testing. *New Genetics and Society,* 35(2), pp. 101–23.

Covolo, L., Rubinelli, S., Ceretti, E. and Gelatti, U. (2015). Internet-Based Direct-to-Consumer Genetic Testing: A Systematic Review. *Journal of Medical Internet Research,* 17(12), e279.

Der Spiegel (2018). Krankenkassen warnen vor Abzocke beim Arzt [Online]. Available at: www.spiegel.de/gesundheit/diagnose/igel-angebote-krankenkassen-kritisieren-selbstzahler-leistungen-als-nutzlos-a-1206029.html (Accessed 24 April 2019).

Deutsches Ärzteblatt (2017). Deutsche sind mit Gesundheitssystem zufrieden [Online]. Available at: www.aerzteblatt.de/nachrichten/83755/Deutsche-sind-mit-Gesundheitssystem-zufrieden (Accessed 24 April 2019).

di Norcia, V. (2012). Consumer Rights. In: Chadwick, R., Callahan, D. and Singer, P., eds., *Encyclopedia of Applied Ethics*, 2nd ed. Amsterdam, Heidelberg: Elsevier, pp. 614–21.

Egglestone C., A. Morris, and O'Brien, A. (2013). Effect of Direct-to-Consumer Genetic Tests on Health Behaviour and Anxiety: A Survey of Consumers and Potential Consumers. *Journal of Genetic Counseling,* 22(5), pp. 565–75.

Einsiedel, E.F. and Geransar, R. (2009). Framing Genetic Risk: Trust and Credibility Markers in Online Direct-to-Consumer Advertising for Genetic Testing. *New Genetics and Society,* 28(4), pp. 339–62.

Encyclopaedia Britannica (2019). 'Caveat Emptor' [Online]. Available at: www.britannica.com/topic/caveat-emptor (Accessed 4 February 2019).

EuroGentest (2019). About Eurogentest [Online]. Available at: www.eurogentest.org/index.php?id=138 (Accessed 24 April 2019).

European Society of Human Genetics (2010). Statement of the ESHG on Direct-to-Consumer Genetic Testing for Health-related Purposes. *European Journal of Human Genetics,* 18(12), pp. 1271–3.

Evans, H. and Wellings, D. (2017). What does the Public think about the NHS? [Online]. Available at: www.kingsfund.org.uk/publications/what-does-public think-about-nhs (Accessed 24 April 2019).

Garstka, H. (2003). Informationelle Selbstbestimmung und Datenschutz. In: Schulzki-Haddouti, C., ed., *Bürgerrechte im Netz.* Wiesbaden: VS Verlag für Sozialwissenschaften, pp. 48–70.

Goldsmith L., Jackson, L., O'Connor, A. and Skirton, H. (2012). Direct-to-Consumer Genomic Testing: Systematic Review of the Literature on User Perspectives. *European Journal of Human Genetics,* 20(8), pp. 811–16.

Gollust, S.E., Gordon, E.S., Zayac, C., Griffin, G., Christman, M.F., Pyeritz, R.E., Wawak, L. and Bernhard, B.A. (2012). Motivations and Perceptions of Early Adopters of Personalized Genomics: Perspectives from Research Participants. *Public Health Genomics,* 15(1), pp. 22–30.

Goold, S.D. (2001). Trust and the Ethics of Health Care Institutions. *Hastings Center Report,* 31(6), pp. 26–33.

Goold, S.D. (2002). Trust, Distrust and Trustworthiness. Lessons from the Field. *Journal of General Internal Medicine,* 17(1), pp. 79–81.

Gordon, E.S., Griffin, G., Wawak, L., Pang, H., Gollust, S.E. and Bernhardt, B.A. (2012). 'It's not like Judgment Day': Public Understanding of and Reactions to Personalized Genomic Risk Information. *Journal of Genetic Counseling,* 21(3), pp. 423–32.

Hall, J.A., Gertz, R., Amato, J. and Pagliari, C. (2017). Transparency of Genetic Testing services for 'Health, Wellness and Lifestyle': Analysis of online prepurchase Information for UK Consumers. *European Journal of Human Genetics,* 25(8), pp. 908–17.

Healthcare for Internationals (2019). Preventive Health Care in the Netherlands [Online]. Available at: https://h4i.nl/2018/11/23/preventive-healthcare-in-the-netherlands/ (Accessed 17 April 2019).

Henneman L., Vermeulen, E., van El, C.G., Claassen, L., Timmermans, D.R. and Cornel, M.C. (2013). Public Attitudes towards Genetic Testing revisited: Comparing Opinions between 2002 and 2010. *European Journal of Human Genetics,* 21(8), pp. 793–9.

Hogarth, S., Javitt, G. and Melzer, D. (2008). The current Landscape for Direct-to-Consumer Genetic Testing: Legal, Ethical, and Policy Issues. *Annual Review of Genomics and Human Genetics,* 9, pp. 161–82.

Kalokairinou, L., Howard, H.C., Slokenberga, S., Fisher, E., Flatscher-Thoni, M., Hartlev M., van Hellemondt, R., Juškevičius, J., Kapelenska-Pregowska, J., Kováč, P., Lovrečić, L., Nys, H., de Paor, A., Phillips, A., Prudil, L., Rial-Sebbag, E., Romeo Casabona, C.M., Sándor, J., Schuster, A., Soini, S., Søvig, K.H., Stoffel, D., Titma, T., Trokanas, T. and Borry, P. (2018). Legislation of Direct-to-Consumer Genetic Testing in Europe: A fragmented regulatory Landscape. *Journal of Community Genetics,* 9(2), pp. 117–32.

Khan, R. and Mittelman, D. (2018). Consumer Genomics will change your Life, whether you get tested or not. *Genome Biology* 19(1), p. 120. doi: 10.1186/s13059-018-1506-1.

Kon, F., Cukier, D., Melo, C., Hazzan, O. and Yuklea, H. (2014). A Panorama of the Israeli Software Startup Ecosystem (March 1, 2014). doi: 10.2139/ssrn.2441157.

Krippendorf, K. (2013). *Content analysis. An introduction to its methodology.* Los Angeles: Sage.

Laestadius, L.I., Rich, J.R. and Auer, P.L. (2017). All your Data (effectively) belong to us: Data Practices among Direct-to-Consumer Genetic Testing Firms. *Genetics in Medicine,* 19(5), pp. 513–20.

Mavroidopoulou, V., Xera, E. and Mollaki V. (2015). Awareness, Attitudes and Perspectives of Direct-to-Consumer Genetic Testing in Greece: a survey of potential consumers. *Journal of Human Genetics,* 60(9), pp. 515–23.

Metcalfe, S.A., Hickerton, C., Savard, J., Terrill, B., Turbitt, E., Gaff, C., Gray, K., Middleton, A., Wilson, B. and Newson, A.J. (2018). Australians' Views on Personal Genomic Testing: Focus Group Findings from the Genioz Study. *European Journal of Human Genetics,* 26, pp. 1101–12.

Oliveri, S., Masiero, M., Arnaboldi, P., Cutica, I., Fioretti, C. and Pravettoni, G. (2016). Health Orientation, Knowledge, and Attitudes toward Genetic Testing and Personalized Genomic Services: Preliminary Data from an Italian Sample. *Biomedical Research International,* Article ID 6824581.

Plothner, M., Klora, M., Rudolph, D. and Graf von der Schulenburg, J.M. (2017). Health-Related Genetic Direct-to-Consumer Tests in the German Setting: The Available Offer and the Potential Implications for a Solidarily Financed Health-Care System. *Public Health Genomics,* 20(4), pp. 203–17.

Raz, A. and Schicktanz, S. (2016). *Comparative Empirical Bioethics. Dilemmas of Genetic Testing and Euthanasia in Israel and Germany.* Berlin: Springer.

Rijksinstituut voor Volksgezondheid en Milieu (2019). Population screening programmes (sic!) [Online]. Available at: www.rivm.nl/en/population-screening-programmes (Accessed 25 April 2019).

Rijksoverheid (2019). Kan ik met een zelftest mijn gezondheid testen? [Online]. Available at: www.rijksoverheid.nl/onderwerpen/bevolkingsonderzoek/vraag-en-antwoord/kan-ik-met-een-zelftest-mijn-gezondheid-testen (Accessed 25 April 2019).

Roberts, J.S., Gornick, M.C., Carere, D.A., Uhlmann, W.R., Ruffin, M.T. and Green, R.C. (2017). Direct-to-Consumer Genetic Testing: User Motivations, Decision Making, and Perceived Utility of Results. *Public Health Genomics,* 20(1), pp. 36–45.

Robertson, R., Appleby, J. and Evans, H. (2018). Public Satisfaction with the NHS and Social Care in 2017 [Online]. Available at: www.kingsfund.org.uk/publications/public-satisfaction-nhs-2017 (Accessed 24 April 2019).

Rosen, B. and Waitzberg, R. (2019). The Israeli Health Care System [Online]. Available at: https://international.commonwealthfund.org/countries/israel/ (Accessed 30 January 2020).

Rosner, G., Rosner, S. and Orr-Urtreger, A. (2009). Genetic Testing in Israel: An Overview. *Annual Review of Genomics and Human Genetics*, 10, pp. 175–92.

Schaper, M., Wöhlke, S. and Schicktanz, S. (2018). 'I would rather have it done by a Doctor' – Laypeople's Perceptions of Direct-to-Consumer Genetic Testing (DTC GT) and its Ethical Implications. *Medicine, Health Care and Philosophy*, 22(1), pp. 31–40.

Schaper, M., and Schicktanz, S. (2018). Medicine, Market and Communication: Ethical Considerations in Regard to Persuasive Communication in Direct-to-Consumer Genetic Testing Services. *BMC Medical Ethics*, 19(56). doi: 10.1186/s12910-018-0292-3

Schicktanz, S. (2012). Epistemische Gerechtigkeit. Sozialempirie und Perspektiven-pluralismus in der Angewandten Ethik. *Deutsche Zeitschrift für Philosophie*, 60(2), pp. 269–83.

Statistisches Bundesamt (2019). Gesundheitsausgaben [Online]. Available at: www.destatis.de/DE/Themen/Gesellschaft-Umwelt/Gesundheit/Gesundheitsausgaben/_inhalt.html (Accessed 24 April 2019).

Sterckx, S. and Cockbain, J. (2014). The UK National Health Service's 'Innovation Agenda': Lessons on Commercialisation and Trust. *Medical Law Review*, 22(2), pp. 221–37.

Stewart, K.F.J., Kokole, D., Wesselius, A., Schols, A., Zeegers, M.P., de Vries H., and van Osch, L.A.D.M. (2018). Factors Associated with Acceptability, Consideration and Intention of Uptake of Direct-To-Consumer Genetic Testing: A Survey Study. *Public Health Genomics*, 21(1–2), pp. 45–52.

The Times of Israel (2019). U.S. Study Finds Israelis Lead the World in Social Media Usage [Online]. Available at: https://blogs.timesofisrael.com/u-s-study-finds-israelis-lead-the-world-in-social-media-usage/ (Accessed 24 April 2019).

Thorlby, R. and Arora, S. (2019). The English Health Care System [Online]. Available at: https://international.commonwealthfund.org/countries/england/ (Accessed 30 January 2020).

van der Schee, E., Braun, B., Calnan, M., Schnee, M. and Groenewegen, P.P. (2007). Public trust in health care: A comparison of Germany, The Netherlands, and England and Wales. *Health Policy*, 81(1), 56. https://doi.org/10.1016/j.healthpol.2006.04.004.

Wammes, J., Jeurissen, P., Westert G. and Tanke, M. (2017). The Dutch Health Care System [Online]. Available at: https://international.commonwealthfund.org/countries/netherlands/ (Accessed 30 January 2020).

Weiner, J.P. (2012). Doctor-Patient Communication in the E-health Era. *Israel Journal of Health Policy*, 1(33). doi: 10.1186/2045-4015-1-33.

Weintraub, K. (2018). Genetics Start-Up Wants to Sequence People's Genomes for Free. *The Scientific American* [Online]. Available at: www.scientificamerican.com/article/genetics-start-up-wants-to-sequence-peoples-genomes-for-free/ (Accessed 12 April 2019).

9 The application of new technologies to improve literacy among the general public and to promote informed decisions in genomics

Serena Oliveri, Renato Mainetti, Ilaria Cutica, Alessandra Gorini and Gabriella Pravettoni

Genetic testing *pret-a-porter* and the new health literacy

The United States and Europe are moving to a prevention-based health care system that will be informed increasingly by genetic perspectives. The significant progress in the identification of genetic variants that can affect human health or can cause disease onset has led to a greater emphasis on genetics in health care. Prior to the sequencing of the human genome through the Human Genome Project, the lay public had little reason to understand human genetics research, as the science was limited and had little impact on most people's daily lives. Indeed, in the past, traditional genetic research focused on conditions relating to a single gene, such as cystic fibrosis and sickle cell disease. Current genomic science, instead, addresses the genetic material as a dynamic system, revealing how genes work and interact with non-genetic factors, including the environment and lifestyle experiences (Portela and Esteller 2010; Sharma et al. 2010). Modern research has revealed genetic associations with the leading causes of mortality, including cancer, diabetes and heart disease (Munger et al. 2007), making genetic research relevant for nearly everyone.

Nevertheless, such advances in human genetics will be useful only when both health care professionals and laypeople have a good understanding of genetic concepts and terminology (Burke et al. 2002). More specifically, laypeople should understand basic human genetics terminology to be able to understand the risk of a genetic condition and have the basic skills to interact with an interdisciplinary team of health professionals. Patients should be able to reconstruct their family history of the disease, construct an appropriate multigenerational pedigree, and understand if they would benefit from a referral to genetic counselling and service.

Unfortunately, considering the increasing complexity of genetic data, reaching this objective is not so easy. The promise of personalized medicine is that it will generate personalized therapies, but it will also generate hard-to-understand personalized risk and benefit information. In particular, there are three main critical concepts that make genetics so complex. The first is called 'pleiotropy' and indicates that each gene contributes to many different traits. The second refers to the fact that each trait is usually the result of the expression of different genes, while the third refers to the evidence that the effect of each genetic variant is related to the characteristics of the environment in which it manifests itself. The situation is further complicated by the fact that there are three different forms of genetic testing: diagnostic, carrier and predictive testing. Diagnostic testing involves identifying current disease states and includes prenatal and newborn screening. Carrier testing determines if an individual carries a certain genetic trait. Finally, predictive testing is used to determine whether a person, who is usually healthy and maybe with a positive family history for a certain disease, has a genetic mutation that will lead to a late-onset disorder.

In the current market of genetic testing, people may perform these tests under the expert guidance of their physicians, or guided by their own curiosity, buying such tests without any kind of prescription or advice; in any case, they should have or receive enough information and knowledge to correctly understand and use the test results as to obtain advantages from such information (Oliveri and Pravettoni 2016). Together with the awareness of these needs, we cannot ignore that current approaches to genetic and genomic education do not adequately prepare the public to understand personal health issues involving genomic medicine, nor to understand media reports about the advancement of genomic research. As a consequence, it is still too common that people are left to interpret genetic results, manage information and make decisions without receiving adequate support from a trained genetic counsellor. This situation might lead to severe misconceptions, inadequate psychological reactions and wrong choices made with negative effects on the individual's life and even on the family members involved.

In the first section of this chapter, we discuss the concepts that are at the basis of an adequate comprehension of genetic risk information, such as health/genetic literacy and numeracy, aspects that the general public would need to acquire to make informed decisions. Then we will focus on the psychological implication related to genetic risk communication, the individual predispositions, and the cognitive aspects that may influence the level of understanding of genetic concepts and the consequent ability to manage genetic results. Finally, we aim to introduce a modern approach to empower people through the use of new technologies and improve their knowledge of genetics. We will focus on the application of serious games for educating laypeople and will present tools developed within the

Mind the Risk project, specifically created to support the players' sense of self-efficacy in managing genetic risk information.

What is known and what needs to be known by laypeople about genetics: health literacy, genomic literacy and numeracy

Low health literacy and numeracy are often associated with markers of disadvantaged status, including poverty, lower educational attainment, belonging to a minority race or ethnicity, and a lack of insurance (National Academies of Sciences, Engineering, and Medicine 2016). All these factors might be associated with various negative medical issues, such as increased risk of illness, increased hospitalization, difficulty with medical information retention and recall, difficulties in accessing the Health System services, prejudices concerning therapies, resistance in the request for clarification, increase in health expenditure, lower participation in health promotion activities, poor compliance with therapeutic indications, and increased morbidity and mortality (Berkman et al. 2011). A further element is that people with low health literacy and numeracy are poorly aware of their family health history, more often than those with high health competence; families do not deeply discuss health difficulties, and physicians engage them less frequently in conversations about genetic topics. Family members who receive positive genetic test results may not disclose them to partners, other family members or insurance companies based on worries and anxiety related to their own identity, as well as a lack of understanding about their family members' risk. Many people wrongly interpret numerical info related to their genetic risk and believe that genes absolutely determine health and disease status. This belief is fed by media translations of genetic science (Parrott et al. 2015). This 'genetic determinism' negatively affect an individual's health-related actions to influence genetic expression, including failure to seek medical care, and contribute to stereotypes and stigmatization.

Health literacy and genomic health literacy

According to the US Department of Health and Human Services and Institute of Medicine, health literacy is 'the degree to which individuals have the capacity to obtain, process, and understand basic health information and services needed to make appropriate health decisions' [https://nnlm.gov/initiatives/topics/health-literacy]. The Institute of Medicine has operationalized health literacy as having the following components: oral literacy (listening and speaking skills), print literacy (reading and writing skills), and numeracy (basic quantitative skills), in addition to cultural and conceptual knowledge (Nielsen-Bohlman et al. 2004). To date, health literacy means much more than reading brochures successfully and managing medical appointments and contents; it is a concept based on the 'empowerment' of people in having access to health information. It represents an observable set of cognitive, individual and social skills required to understand and use information as appropriately as possible in terms of health-related decisions. These skills enable individuals to obtain, understand and use information to make decisions and take actions that will have an impact on their health status. In terms of the breadth, dynamism and constant changes that characterize the health system, it is actually difficult to define the boundaries of a concept such as health literacy as it is in continuous expansion. In absolute terms, we should distinguish between those who have basic skills that enable them to access, understand and use information for health purposes and those who do not. In relative terms, we assess the skill differences between those who are able to apply more advanced cognitive and literacy skills to perform relatively challenging tasks for health purposes and those who cannot.

In November 2011, during the meeting that was held by the National Human Genome Research Institute, *Genomic Health Literacy* was defined as 'the knowledge of basic genetics and genomics concepts and processes needed to build conceptual understanding, and the necessary mathematical knowledge to support this comprehension' (Hurle et al. 2013). Unfortunately, the public's level of understanding of basic biology and mathematics (including probability theory, statistics and risk) is still very poor (Lanie et al. 2004; Lea et al. 2011). For instance, it has been found that a relevant number of individuals have a vague idea that genes are inherited, without knowing how genes effectively work; other people understand that genetic mutations can cause a disease and some of them also know that the environment impacts the expression of the disease in each individual (Smith et al. 2014). Also, most people do not give a primary role to the genes in the causation of all human characteristics (Parrott et al. 2003), and do not believe that an individual carrying a gene variant will certainly develop an associated health condition (Bates et al. 2003). Moreover, it has been found that despite the fact that most people have a limited understanding of what genes are, how they are inherited and where they are located in the body, they believe that they well understand information about genes anyway (Lanie et al. 2004). Mesters and colleagues (2005) found a similar overestimation of understanding among the participants in their focus group discussions: although participants discussed genes and DNA as strictly related to cancer, their understanding of these concepts was limited to a strong familiarity with the words and not with the concept itself.

The paucity in lay knowledge about genetics, health and risk persists even among individuals with direct experience with single-gene disorders. Family members may know about the effects of a genetic disorder due to exposure to an affected relative, but frequently they ignore their own risk for the same condition, the fact that they can pass it on to the future offspring, and they are not aware of the carrier frequency in the general population. Most members of the public have not received, and probably will never receive, formal genetics education (Condit 2010a), and this creates gaps in understanding genetic concepts. Even individuals with formal education in biology and genetics, including undergraduate and medical students nearing graduation, as well as health educators and medical experts, often lack sufficient understanding of genetics (Baars et al. 2005; Bowling et al. 2008; Chen and Goodson 2007).

Health numeracy

Difficulties in comprehension are mainly due to the same nature of genetic risk information (probabilistic, uncertain and linked to abstract concepts) and to aspects related to the individual (general beliefs about health, the perception of the disease and its controllability, past experiences and the tendency to use heuristics). The main aspect of risk communication is the abstract nature of its quantitative format and of the associated statistical information (a probability distribution or frequency). Health numeracy is defined as 'the degree to which individuals have the capacity to access, process, interpret, communicate, and act on numerical, quantitative, graphical, biostatistical, probabilistic health information needed to make effective health decisions' (Goldbeck et al. 2005, p. 375). Numeracy is important because it can influence how and what information is processed and understood in decision-making (Ancker and Kaufman 2007; Fagerlin et al. 2007).

Based on the National Adult Literacy Survey, almost half of the general population has difficulty with relatively simple numeric tasks such as calculating (using a calculator) the difference between a regular price and a sales price (Kutner et al. 2006). Having fewer numeric skills is associated with lower comprehension and less use of health information. Many patients cannot perform the basic numeric tasks required to function in the current health care environment. For example, 26 per cent of the participants in one study were unable to understand information about when an appointment was scheduled (Williams et al. 1995) and studies found that even highly educated people incorrectly answer straightforward questions about risk magnitudes (for example: Which represents the larger risk: 1 per cent, 5 per cent or 10 per cent?) (Lipkus et al. 2001; Peters et al. 2007a, 2007b).

Understanding numeric information about health risk estimation in real-life situations is even more complex. Indeed, numerical information, associated with risk estimation, can be expressed essentially in two formats: the percentage format and the frequency format. The percentage format is the most difficult numeric format to understand, as it is not always clear to what risk the percentage refers to (Thomson et al. 2005; Timmermans et al. 2008). To give an example of the difficulties in understanding the percentage format, let us consider the following sentence: 'Mammography reduces the risk of breast cancer by 25 per cent.' The meaning of this sentence could be ambiguous. What does that 25 per cent refer to? The sentence provides an estimate of relative risk, that is the risk associated with all those women who undergo mammography, but it is not known what the absolute risk is, nor the reference time frame. A better way to present the same information could be: 'of 1000 women who do not undergo mammography, four will die of breast cancer in the next ten years, whereas for 1000 women who undergo mammography, three will die' (Thomson et al. 2005). The frequencies, on the other hand, are easier to understand and to imagine than the percentages, particularly if the risk is expressed as a natural frequency, so it is clear what the reference class is (the relationship between numerator and denominator) (Timmermans et al. 2008). For example, saying that '20 out of 100 women who have been found to have a *BRCA2* gene mutation will develop ovarian cancer later in life' is clear and understandable and leaves no room for misunderstanding. Although it has been shown that the level of numeracy does not affect the decision to do the *BRCA* predictive test, unlike emotional involvement (Shatz et al. 2015), having the ability to understand a numerical value also means having more chances that the decision is taken in an informed manner.

Individuals who have low ability to manipulate and understand the information in a numerical format, tend to rely their decision upon non-numeric information, and generally exhibit greater susceptibility to heuristics and biases (Reyna et al. 2015). Furthermore, people with low numeracy have different needs in terms of decision aids: the comprehension and the quality of their health decisions might depend on the interaction between their level of numeracy and the format with which the information is provided (Peters et al. 2007). Since the individual level of numeracy might constitute a valid indicator of the potential level of understanding, it is necessary for a clinician to identify the most useful channels and formats for communicating genetic risk information (Lea et al. 2011).

How people elaborate genetic knowledge of disease predisposition

Psychological aspects

The awareness of being carrier of a mutation that predisposes to a specific disease is paramount for taking preventive actions, and to promote early detection of diseases and treatments in most cases. However, there are many possible psychological reactions and impacts on people's daily lives related to receiving genetic risk information. Most studies on the psychological impact of genetic tests reported 'harmful' reactions, such as anxiety, distress and depression after receiving genetic results (Oliveri 2016a), and found that DNA-based disease risk information had little or no effect on health-related behaviours (Heshka et al. 2008; Hollands et al. 2016).

However, this does not mean that genetic testing is not useful to bring an added value for people's knowledge, awareness and actual change in health habits and decisions. We should indeed consider that genetic testing impact also depends on how people perceive their risk, severity and controllability related to specific categories of disease (Cameron and Muller 2009; Wang et al. 2009; Wade et al. 2012). Moreover, the genetic testing impact depends on the level of predictability or the same nature of the diseases (from monogenic to genetic susceptibility factors), and on the presence/absence of treatments (Cameron and Muller 2009).

When considering genetic testing for establishing the risk of a specific disease, we are talking about searching for genes/mutations that are not able, on their own, to determine the disease: in other words, the genetic analysis shows only a general susceptibility to getting sick. The so-called multifactorial diseases are caused by the expression of more than one gene and their onset also depends on a particular interaction between genes and the environment.

A recent review (of the last 20 years of contributions) about the psychological impact of communicating genetic test results (Oliveri et al. 2018a), reported that there is no significant increase in distress levels or an adverse impact on the quality of life in subjects who undergo a genetic test for cardiovascular disease, cancer or Alzheimer's disease. The psychological distress seems to be usually related to a full-blown clinical condition in addition to a positive genetic result, thus individuals who already have physical symptoms tend to be more highly motivated to change their lifestyles after results. Individuals with full-blown symptoms are also described as motivated to gather health-related information, including genetic risk information, in order to manage their risk of developing the disease. Different results emerged for Huntington disease (HD), where reactions such as depressive symptoms, suicidal ideations and hopelessness characterize gene carriers. Negative reactions may be understandable in the light of the certainty these people have to develop HD in the future, the perception of something uncontrollable and fatal, alongside the complete absence of valid therapies and inevitable cognitive decline. Overall, the review results show that people tend to consider genetic tests as valid information to take important preventive decisions, despite the personal differences in risk perception, worry and other psychological reactions or the differences in disease controllability and existing therapies to manage the illness.

When an individual undergoes a genetic test, there are psychological implications for family members. If the mutation is not detected by the genetic analysis, the individual's emotional reaction will generally be of great relief and profound contentment. This relief, on the other hand, can dangerously translate into behaviours that violate the standard guidelines for health prevention. On the contrary, when a genetic susceptibility is detected, the psychological reaction of the whole family may be based on the type of disease investigated and the percentage of risk. The proband may feel guilty to be the 'author' of the future illness of loved ones. Feelings of pain are also possible, due to profound sorrow and anger towards oneself and/or others.

Family members, who discover that they have a real chance to get sick, come face-to-face with an unpleasant reality that they have to manage and keep under control. In some cases, they may make drastic and difficult choices, such as undergoing a preventive surgery (e.g. the prophylactic mastectomy after *BRCA* mutation detection). Furthermore, they sometimes have to deal with life choices that would have been postponed, such as a planned pregnancy. The psychological consequences are the uncertainty and the anxiety of not knowing whether or when the pathology will develop, the fear of suffering and of not being able to defeat the disease.

The psychological effects on family members also depend on past experience, when they may have already lived with affected relatives. A meta-ethnographic review of qualitative research studies found that individuals' experiences of family members' illnesses and personal models of disease causation influenced their perceived vulnerability (Walter et al. 2004). Although one may rationally know that there is a profound difference between their loved one's clinical experience and their situation, the experience of the other is an anchor from which to evaluate one's own situation.

Cognitive aspects: health beliefs, heuristics and biases

Individuals' perceptions, beliefs and past experiences may influence the level of understanding about genetic concepts. The public's understanding of genetics ranges from beliefs that genes make them susceptible to various conditions to beliefs that genes absolutely determine health status (Parrott et al. 2012). The public tends to attribute genetic origins for obesity (Ogden and Flanagan 2008) and cancer (Molster et al. 2009), but often fails to attribute genetic origins for mental health conditions, including depression, schizophrenia and bipolar disorder.

The belief that genes are primarily responsible for causing disease may limit a person's belief in the ability to improve health through personal action, such as smoking cessation, physical activity or dietary changes. It may also hinder the individual's perceived efficacy of medical interventions: personal beliefs are associated with fatalism, and fatalism contributes to the failure to seek health care services (Peek et al. 2008; Shen et al. 2009).

When an individual, even without a pre-test consultation, decides to undergo the predictive genetic test, he/she makes a choice based on a cost/benefit evaluation: in particular, he/she makes a prediction on the possible positive and negative consequences of carrying out the test. This decision-making process represents a very frequent judge ment heuristic in health decision-making, which is often prone to distortions about the forecast assessment of a future emotional state (affective forecasting heuristic).

Often, individuals tend to overestimate the impact that future (especially negative) emotional events might have on their psychophysical well-being (Shatz et al. 2015): such a tendency is called impact bias. Impact bias, in anticipation of future emotional states, often leads to an overestimation of the perceived risk, both by the patient (affective forecasting bias) and by the doctor, who anticipates his own emotional state (emphatic forecasting bias), characterized by excessive risk aversion. Furthermore, the future anticipation of an excessive emotional intensity referred to one's own health conditions, is higher in healthy subjects than in actually affected subjects, who, for example, have already been diagnosed with a life-threatening condition (disability paradox). In general, patients tend to overestimate the emotional impact of a positive result (detected mutation) of a genetic test. Evidence in literature reveals that the individual's level of stress peaks immediately after having received the genetic results for an increased risk of cancer, but the stress returns to normal levels over time (Lerman et al. 2002; Peters et al. 2013). The impact bias might partially explain the underutilization of predictive genetic tests by family members of a proband already classified as being 'at-risk'.

The literature suggests that the individual's expectation for future emotional states is influenced by the evaluation of a current stressful situation (first evaluation) and by the perceived ability to react to it (second evaluation) (O'Neill et al. 2015). Let's consider, for instance, the case of a woman with a family history of breast and ovarian cancer: she might decide to undergo the *BRCA* 1/2 predictive genetic test because of her personal risk and the risk for her offspring to develop cancer in the future. Therefore, the decision to take the test for this adult woman could go along with her perceived ability to face a positive result as well as with her expectation about the emotional stress and reactions of her family members (e.g. her daughter) (O'Neill et al. 2015).

In conclusion, the interpretation of risk information is a subjective process: the individual interprets his own risk of cancer through the use of heuristics, through the mediation of his/her beliefs on the inheritance for cancer, the family experience, the need for control, social comparison and subjective motivations (Vos et al. 2012).

Improvement of general public literacy and 'empowerment' of people in genetics through the new technologies

Shared decision-making is now a great challenge in the medical field, considering the growing role of the patient in the medical process. The current trend is to define the decision-making process in terms of 'patient-centred', to indicate the increased responsibility of the patient in the interaction with a doctor and in the evaluation of costs and benefits of a choice, when it is not quantifiable *a priori* (Pravettoni et al. 2006). The medical decision is then taken together with the patient, respecting his values and preferences (Reyna et al. 2015). The main concern is that the patient could not have an adequate level of literacy about the topic discussed during the communication with the physicians, to make a real informed medical decision. As we have seen in the previous sections, the new possibilities given by progress in genetic research and testing does not go hand-in-hand with the lay people's knowledge in this field (Klitzman 2010; Condit 2010b; Oliveri 2016b).

Currently, numerous initiatives are aimed at improving general public literacy in genetics and shared decision-making (Machluf and Yarden 2013; van Mil et al. 2010), such as the Genetic Literacy, Education and Empowerment Initiative (GLEE), a U.S. national campaign to enhance the genomic literacy of the population. It focuses on three target audiences: K-16 students, the general public, and health care providers. In Europe, several websites that provide online courses, lessons and information on genetics (www.eurogentest.org/index.php?id=894) have been created by educational centres or provided by private companies for the general public (but can be accessed by health professionals too).

The most famous company delivering Direct to Consumer Genetic Testing (DTC-GT), 23andMe, launched its Education Program after a survey conducted in 2013 on its clients. Results from this study demonstrated that more than 80 per cent of clients had difficulties in understanding some basic concepts about genetics. Moreover, 23andMe's popular video series, 'Genetics 101', is a mainstay for anyone wanting to learn more about science.

The general public includes absolutely heterogeneous ages, cultures and habits. In order to raise an adequate interest in genetics, it is necessary to find a strategy and the right instruments. As allies, we now have a variety of tools in several domains, from books to movies, from videos to interactive apps, and finally video games. The aim is to be able to skilfully produce interesting and captivating content for the population. This approach seems a possible way to engage people on topics concerning genetic risk for common disorders.

Application of new technologies to foster genetic knowledge and education

The National Human Genome Research Institute Meeting Report (Hurle et al. 2013) promotes the use of new media and new technologies to disseminate genomic information to effectively engage the public. Some of the most comprehensive English-language sites, useful for students to learn genetics and equally useful for teachers, were created by the University of Utah and the Genetic Science Learning Center (https://teach.genetics.utah.edu/). Through the material published on these sites, it is possible to address and study many issues related to genetics, biosciences and health. For over 20 years, these sites have been providing free multimedia educational material, specifically designed for students and teachers. They are among the most used scientific sites, with tens of millions of visitors every year. This approach is a good example of how to use multimedia interactivity to make learning attractive and functional. The material is produced by scientists who have succeeded in generating texts that are easy and clear to understand, and rich in content at the same time. The site's collection provides users with material for learning about genetics in various forms, such as texts accompanied by images and videos, but among all these 'classical modalities' the interactive content really stands out. Through simulations, users learn by doing, following tutorials and training with different aspects of genetics. The most basic theoretical parts of these e-learning tools introduce the terminology and explain the basic concepts of genetics, whereas other sections allow learning to match up chromosomes in a karyotype. The educational material in these web sites lacks a very important aspect: a 'playful' component that is able to completely engage laypeople and the general public in general. People might be not so interested in having a detailed and rich description of all basic genetic mechanisms and therefore do not find it interesting to spend time on this kind of educational content. The Serious Games (SG) approach seems to find a solution to these aspects: SG is a highly interactive medium that could be adapted to the pace of the user, use multiple visual and auditory modes to present information, provide immediate feedback with tailored instructions and encourage experiential learning promoting self-efficacy, goal setting and cooperation. Evidence shows that SG is efficient in promoting education, training, health promotion and socialization (Susi et al. 2007). Serious Games can attract the general public through game mechanics and challenges, can provide people enough information to understand complex issues and simulate real-life cases. Every day, millions of people play video games that recreate situations of everyday life that are sometimes appropriately modified to adapt to the constraints dictated by the game design. The video game is substantially a didactic device, a machine which encourages the learning process by making mistakes, by trial and error sessions: making mistakes is part of the game.

In the last few years, there have been different attempts to improve individuals' genetic knowledge using SG (e.g. Touching Triton (Loftin et al. 2016), Geniverse [https://geniverse-lab.concord.org], DNA Roulette [http://gel.msu.edu/DNAroulette/more.html]). Unfortunately, current genetic games are mostly addressed to trainees in biology courses and medical practitioners (geneticists); they use a very technical language or they focus on certain aspects of genetics, such as the probabilistic nature of genomic, neglecting the complexity of managing such information in the daily life.

Learning through Serious Games and the endorsement of 'self-efficacy' in genetics

Learning can be viewed both as information acquisition and as knowledge construction. Individuals' ability to properly act in a given situation depends both on the knowledge they have acquired, and on their ability to 'transfer' that knowledge in the situation they are actually in (Arnab et al. 2012). In his book, Mayer (2014) highlights the importance in understanding how the human mind works to be able to introduce the best elements in designing computer games for learning. Mayer (2014) claims that games for learning must have a specific characteristic in order to be defined as such: they must cause a measurable change in the player's knowledge or cognitive abilities, indeed 'the gold standard in game effectiveness research is to measure a change in what the learner knows or can do'. Games for learning can be based on games and simulations that can be played both in reality and through the use of a computer. In SGs, the 'serious' part deals with trying to make players literate with the subject of training, helping them to mentally organize information through a consistent cognitive representation and integrate them with previous knowledge related to the same subject/topic.

According to Mayer (2014) and other authors (Bergey et al. 2015; Starks 2014), an element that could be used to evaluate if people are going to learn better in games for learning is self-efficacy. Perceived self-efficacy is defined as 'people's beliefs about their capabilities to produce designated levels of performance that exercise influence over events that affect their lives. Self-efficacy beliefs determine how people feel, think, motivate themselves and behave' (Bandura 1994). SGs could act on people's self-efficacy through two different modalities. First, the learning task is hidden within a game, where elements stimulating self-efficacy could be implemented without overwhelming the player (Kato 2012). In this way, the real/main task (learning task) is not explicitly presented to the player and the player is less critical about his/her performance. Second, if the game manages to achieve its purpose, which is to transfer knowledge to the player, at the end of the game session the player will feel more confident about the topic and efficient in managing future tasks dealing with it. This last aspect is really important, especially in circumstances where the training process does not end but must continue over time due to the increasing complexity of the topic itself, such as occurs in the genetic field.

The Mind the Risk project: explore the use of Serious Games for public education in genetics

Our vision encourages the application of new technologies, SGs in particular, to enhance laypeople's self-efficacy in managing genetic risk information. Part of our contribution to the *Mind the Risk* project consisted in the identification (through deep scientific research in literature, discussions with health professionals and scientific popularisers) of those essential genetic notions which are the pillars of genetics and need to constitute the ground of public literacy in genetics. We furthermore discussed the best way to explain, teach and convey complex genetic concepts and mechanisms in an entertaining and simpler way. We did not forget, however, the emotional and experiential part which goes along with the experience of undergoing a genetic test and discovering to have a significant mutation, affecting one's own health and the family (Oliveri et al. 2015).

Our aim was to reproduce this 'experiential/real-life dimension' in designing new Serious Games for learning in genetics. Three different Serious Games were fully designed, developed and tested for this project. Two mini-games have been created with the aim of introducing different basics concepts on inheritance and mutations: HerediRabbit and MutanTetris (Oliveri et al. 2018b). An adventure game 'Gene Adventure the world of tomorrow' (Oliveri et al. 2018b) has been designed to make people experience, in a simulated life, the management of complex information such as genetic risk, and develop a deeper understanding of genetic mechanisms and the consequences of behaviours on gene expression.

According to the social cognitive theories, which recognizes the importance of self-efficacy as a prerequisite of people change in behaviour (Bandura 2005), we believe that the Serious Game Gene Adventure-wt can support players' sense of self-efficacy and agency in the management of genetic risk information. It indeed proposes simulated dynamics in daily life that the players can independently manage through a functional trial and error process or by stimulating the processes of identification with the avatar of the game.

Take home messages

- Progresses in genetics will generate personalized health care pathways but will also generate hard-to-understand personalized risk and benefit information.
- Lay people should understand basic human genetics terminology, their risk of a genetic condition, and should have the basic skills to interact with an interdisciplinary team of health professionals and make informed decisions.
- The interpretation of risk information is a subjective process: the individual interprets his own risk of cancer, through the use of heuristics, through the mediation of his/her beliefs on the inheritance of mutation and cancer in general, the family experience, the need for control, social comparison and subjective motivations.
- Serious Games (SG) and new technologies are a valid instrument to increase genetic literacy in the general public. They are highly interactive medium, adapted to the pace of the user, use multiple visual and auditory modes to present information, provide immediate feedback with tailored instructions and encourage experiential learning promoting self-efficacy in genetics.

References

Ancker J. and Kaufman, D. (2007). Rethinking health numeracy: a multidisciplinary literature review. *J Am Med Inform Assoc*, 14: 713–21.

Arnab, S., Berta, R., Earp, J., de Sara, F., Popescu, M., Romero, M., Stanescu, I. and Usart, M. (2012). Framing the adoption of serious games in formal education. *Electronic Journal of e-Learning.*

Baars, M., Henneman, L. and Kate, L.P. (2005). Deficiency of knowledge of genetics and genetic tests among general practitioners, gynecologists, and pediatricians: A global problem. *Genetics in Medicine*, 7, 605–10.

Bandura, A. (1994). Self-efficacy. In V.S. Ramachaudran (Ed.), *Encyclopedia of Human Behavior* (Vol. 4, pp. 71–81). New York: Academic Press. (Reprinted in H. Friedman (Ed.), *Encyclopedia of Mental Health*. San Diego: Academic Press, 1998).

Bandura, A. (2005). The primacy of self-regulation in health promotion. *Applied Psychology*, 54(2), pp. 245–54.

Bates, B.R., Templeton, A., Achter, P.J., Harris, T.M. and Condit, C.M. (2003). What does 'a gene for heart disease' mean? A focus group study of public understanding of genetic risk factors. *Am J Med Genet A*, 119A, 156–61.

Bergey, B.W., Ketelhut, D.J., Liang, S., Natarajan, U. and Karakus, M. (2015). Scientific inquiry self-efficacy and computer game self-efficacy as predictors and outcomes of middle school boys' and girls' performance in a science assessment in a virtual environment. *Journal of Science education and Technology*, 24(5), 696–708.

Berkman, N.D., Sheridan, S.L., Donahue, K.E., Halpern, D.J. and Crotty, K. (2011). Low health literacy and health outcomes: an updated systematic review. *Annals of internal medicine*, 155(2), 97–107.

Bowling, B.V., Acra, E.E., Wang, L., Myers, M.F., Dean, G.E., Markle, G.C., Moskalik C.L. and Huether, C. (2008). Development and evaluation of a genetics literacy assessment instrument for undergraduates. *Genetics*, 178, 15–22.

Burke, W., Atkins, D., Gwinn, M., Guttmacher, A., Haddow, J., Lau, J., Palomaki, G., Press, N., Richards, C.S., Wideroff, L. and Wiesner, G.L. (2002). Genetic test evaluation: Information needs of clinicians, policy makers, and the public. *Am J Epidemiol,* 156(4), 311–18.

Cameron, L.D. and Muller, C. (2009). Psychosocial aspects of genetic testing. *Curr. Opin. Psychiatry,* 22, 218–23. doi: 10.1097/YCO.0b013e3283252d80

Chen, L.S. and Goodson, P. (2007). Public health genomics knowledge and attitudes: A survey of public health educators in the United States. *Genetics in Medicine*, 9, 496–503.

Condit, C.M. (2010a). Public Understandings of Genetics and Health. *Clinical Genetics*, 77, 1–9.

Condit, C.M. (2010b). Public Attitudes and Beliefs About Genetics. *Annual Review of Genomics and Human Genetics*, 11(1), 339–359.

Fagerlin, A., Ubel, P.A., Smith, D.M. and Zikmund-Fisher, B.J. (2007). Making numbers matter: present and future research in risk communication. *Am J Health Behav.*, 31(suppl 1), 47–56.

Goldbeck, A.L., Ahlers-Smith, C.R. and Paschal, A.M. (2005). A definition and operational framework for health numeracy. *Am J Prev Med.*, 29, 375–6.

Heshka, J.T., Palleschi, C., Howley, H., Wilson, B. and Wells, P.S. (2008). A systematic review of perceived risks, psychological and behavioral impacts of genetic testing. *Genet. Med.*, 10, 19–32. doi: 10.1097/GIM.0b013e31815f524f

Hollands, G.J., French, D.P., Griffin, S.J., Prevost, A.T., Sutton S., King S. and Marteau T.M. (2016). The impact of communicating genetic risks of disease on risk-reducing health behaviour: Systematic review with meta-analysis. *BMJ*, 352. 10.1136/bmj.i1102

Hurle, B., Citrin, T., Jenkins, J.F., Kaphingst, K.A., Lamb, N., Roseman, J.E. and Bonham, V.L. (2013). What does it mean to be genomically literate? National Human Genome Research Institute meeting report.

Kato, P.M. (2012). Putting self-efficacy theory into serious games – Pamela M. Kato, EdM PhD. https://pamkato.com/2012/02/22/putting-self-efficacy-theory-into-serious-games/.

Klitzman R.L. (2010). Misunderstandings concerning genetics among patients confronting genetic disease. *Journal of Genetic Counseling*, 19(5), 430–46.

Kutner, M., Greenberg, E., Jin, Y. and Paulsen, C. (2006). The health literacy of America's adults: Results from the 2003 National Assessment of Adult Literacy. Washington, DC: National Center for Educational Statistics.

Lanie, A.D., Jayaratne, T.E., Sheldon, J.P., Kardia, S.L., Anderson, E.S., Feldbaum, M. and Petty, E.M. (2004). Exploring the public understanding of basic genetic concepts. *Journal of genetic counseling*, 13(4), 305–20.

Lea, D.H., Kaphingst, K.A., Bowen, D., Lipkus, I. and Hadley, D.W. (2011) Communicating genetic and genomic information: health literacy and numeracy considerations. *Public Health Genomics*, 14(4–5), 279–89.

Lerman, C., Croyle, R.T., Tercyak, K.P. and Hamann, H. (2002). Genetic Testing: psychological aspects and Implications. *Journal of Counseling and Clinical Psychology*, 70(3), 784–97.

Lipkus, I.M., Samsa, G. and Rimer, B.K. (2001). General performance on a numeracy scale among highly educated samples. *Medical decision making*, 21(1), 37–44.

Loftin, M., East, K., Hott, A. and Lamb, N. (2016). 'Touching Triton': Building Student Understanding of Complex Disease Risk. *American Biology Teacher*, 78(1), 15–21.

Machluf, Y. and Yarden, A. (2013). Integrating bioinformatics into senior high school: design principles and implications. *Briefings in Bioinformatics*, 14(5), 648–60.

Mayer, R.E. (2014). Computer games for learning: An evidence-based approach. Cambridge, UK: MIT Press.

Mesters, I., Ausems, A. and DeVries, H. (2005). General public's knowledge, interest and information needs related to genetic cancer: an exploratory study. *Eur J Cancer Prev*, 14, 69–75.

Molster, C.T., Samanek, A. and O'Leary, P. (2009). Australian study on public health knowledge of human genetics and health. *Public Health Genomics*, 12, 84–91.

Munger, K.M., Gill, C.J., Ormond, K.E. and Kirschner, K.L. (2007). The next exclusion debate: Assessing technology, ethics, and intellectual disability after the Human Genome Project. *Mental Retardation and Developmental Disabilities*, 13, 121–8.

National Academies of Sciences, Engineering, and Medicine (2016). *Relevance of health literacy to precision medicine: Proceedings of a workshop*. National Academies Press.

Nielsen-Bohlman, L., Panzer, A.M. and Kindig, D.A. (eds) (2004). *Health Literacy: A Prescription to End Confusion*. Washington: National Academies Press.

O'Neill, S.C., Mays, D., Partenaude, A.F., Garber, J.E., DeMarco, T.A., Peshkin, B.N., Schneider, K.A. and Tercyak, K.P. (2015). Women's concerns about the emotional impacts of awareness of heritable breast cancer risk and its implications for their children. *Journal Community Genetic*, 6, 55–62.

Ogden, J. and Flanagan, Z. (2008). Beliefs about the causes and solutions to obesity: A comparison of GPs and lay people. *Patient Education and Counseling*, 71, 72–8.

Oliveri, S. and Pravettoni, G. (2016). The disclosure of direct-to-consumer genetic testing: sounding out the psychological perspective of consumers. *Biology and Medicine*, 8(5), 1.

Oliveri, S., Ferrari, F., Manfrinati, A. and Pravettoni, G. (2018a). A Systematic Review of the Psychological Implications of Genetic Testing: A Comparative Analysis Among Cardiovascular, Neurodegenerative and Cancer Diseases. *Frontiers in genetics*, 9.

Oliveri, S., Howard, H.C., Renzi, C., Hansson, M.G. and Pravettoni, G. (2016a). Anxiety delivered direct-to-consumer: are we asking the right questions about the impacts of DTC genetic testing? *Journal of Medical Genetics*, 53(12), 798–99.

Oliveri, S., Mainetti, R., Gorini, A., Cutica, I., Candiani, G., Borghese, N.A. and Pravettoni, G. (2018b). Serious Games for Improving Genetic Literacy and Genetic Risk Awareness in the General Public: Protocol for a Randomized Controlled Trial. *JMIR research protocols*, 7(12), e189.

Oliveri, S., Masiero, M., Arnaboldi, P., Cutica, I., Fioretti, C. and Pravettoni, G. (2016b). Health orientation, knowledge, and attitudes toward genetic testing and personalized genomic services: Preliminary data from an Italian sample. *BioMed research international* 2016:6824581. doi: 10.1155/2016/6824581

Oliveri, S., Renzi, C., Masiero, M. and Pravettoni, G. (2015). Living at risk: factors that affect the experience of direct-to-consumer genetic testing. *Mayo Clinic Proceedings* 90(10), 1323–6.

Parrott, R.L., Silk, K.J. and Condit C. (2003). Diversity in lay perceptions of the sources of human traits: genes, environments, and personal behaviors. *Soc Sci Med*, 56, 1099–1109.

Parrott, R., Kahl, M.L., Ndiaye, K. and Traeder, T. (2012). Health communication, genetic determinism, and perceived control: The roles of beliefs about susceptibility and severity versus disease essentialism. *Journal of health communication*, 17(7), 762–78.

Parrott, R.L., Worthington, A.K., Smith, R.A. and Chadwick, A.E. (2015). Communicating about genes, health, and risk. In *Oxford Research Encyclopedia of Communication*. Oxford University Press. doi: 10.1093/acrefore/97801

Peek, M.E., Sayad, J.V. and Markwardt, R. (2008). Fear, fatalism and breast cancer screening in low-income African-American women: The role of clinicians and the health care system. *Journal of General Internal Medicine*, 23(11), 1847–53.

Peters, E., Dieckmann, N., Dixon, A., Hibbard, J.H. and Mertz, C.K. (2007a). Less is more in presenting quality information to consumers. *Medical Care Research and Review*, 64(2), 169–90.

Peters, E., Hibbard, J., Slovic, P. and Dieckmann, N. (2007b). Numeracy skill and the communication, comprehension, and use of risk-benefit information. *Health Affairs*, 26(3), 741–48.

Peters, S.A., Laham, S.M. and Pachter, N. (2013). The future in clinical genetics: affective forecasting biases in patient and clinician decision making. *Clinical Genetics*, doi: 10.1111/cge.12255, 1–6.

Portela, A. and Esteller, M. (2010). Epigenetic modifications and human disease. *Nat Biotechnol*, 28(10), 1057–68.

Pravettoni, G., Lucchiari, C. and Vago, G. (2006). Quality of life and shared decisions in patients with high grade gliomas. In *Better decisions, better care: advancing decision support to improve health care* (pp. 14–15). Boston: SAGE Publications.

Reyna, V.F., Nelson, W.L., Han, P.K. and Pignone, M.P. (2015). Decision making and cancer. *American Psychologist*, 70(2), 105–18.

Sharma, S., Kelly, T.K. and Jones, P.A. (2010). Epigenetics in cancer. *Carcinogenesis*, 31(1), 27–36.

Shatz, M.T., Hanoch, Y., Katz, B.A., Doniger, G.M. and Ozanne, E.M. (2015). Willingness to test for BRCA 1/2 in high risk women: Influenced by perception and family experience, rather than by objective or subjective numeracy? *Judgement and Decision Making*, 10(4), 386–99.

Shen, L., Condit, C. and Wright, L. (2009). The psychometric property and validation of a fatalism scale. *Psychology & Health*, 24, 597–613.

Smith, R.A., Greenberg, M. and Parrott, R.L. (2014) Segmenting by risk perceptions: Predicting young adults' genetic-belief profiles with health and opinion-leader covariates. *Health Communication*, 29, 483–93.

Starks, K. (2014). Cognitive behavioral game design: a unified model for designing serious games. *Frontiers in psychology*, 5, 28.

Susi, T., Johannesson, M. and Backlund, P. (2007) 'Serious Games - An Overview'. Technical Report HS-IKI-TR-07-001, School of Humanities and Informatics, University of Skövde, Sweden.

Thomson, R., Edwards, A. and Grey, J. (2005). Risk communication in the clinical consultation. *Clinical Medicine*, 5, 465–469.

Timmermans, D., Ockuhusen-Vermey, C.F. and Henneman, L. (2008). Presenting Health Information in different formats: The effect on participants' cognitive and emotional evaluation and decisions. *Patient Education and Counseling*, 73, 443–47.

Van Mil, M.H., Boerwinkel, D.J., Buizer-Voskamp, J.E., Speksnijder, A. and Waarlo, A.J. (2010). Genomics education in practice: evaluation of a mobile lab design. *Biochemistry and Molecular Biology Education*, 38(4), 224–9.

Vos, J., Oosterwijk, J., Gomez-Garcia, E., Menko, F., Collee, M., Van Asperen, C.J., Jansen, A., Stiggelbout, A.M. and Tibben, A. (2012). Exploring the short-term impact of DNA-testing in breast cancer patients: The counselee's perception matters, but the actual BRCA 1/2 result does not. *Patient Education and Counseling*, 86, 239–51.

Wade, C.H., Shiloh, S., Woolford, S.W., Roberts, J.S., Alford, S.H., Marteau, T.M. and Biesecker, B.B. (2012). Modelling decisions to undergo genetic testing for susceptibility to common health conditions: an ancillary study of the Multiplex Initiative. *Psychol. Health*, 27, 430–44.

Walter, F.M., Emery, J., Braithwaite, D. and Marteau, T.M. (2004). Lay understanding of familial risk of common chronic disease: a systematic review and synthesis of qualitative research. *Ann Fam Med.*, 2, 583–94.

Wang, C., O'Neill, S.M., Rothrock, N., Gramling, R., Sen, A., Acheson, L.S., Rubinstein, W.S., Nease, D.E. and Ruffin, M.T. (Family Healthware Impact Trial (FHITr) group) (2009). Comparison of risk perceptions and beliefs across common chronic diseases. *Prev. Med. (Baltim)*, 48, 197–202. doi: 10.1016/j.ypmed.2008.11.008

Williams, M.V., Parker, R.M., Baker, D.W., Parikh, N.S., Pitkin, K., Coates, W.C. and Nurss, J.R. (1995). Inadequate Functional Health Literacy among Patients at Two Public Hospitals. *Journal of the American Medical Association*, 274(21), 1677–82.

10 Algorithms as emerging policy tools in genomic medicine

Opportunities and challenges ahead

Frederic Bouder

The emergence of algorithms in the medical field

Artificial intelligence (AI) has become omnipresent in our daily lives, from search and recommendation engines (e.g. Google search), to personal digital assistants and smart home appliances. It is expected to become increasingly important for medical decisions such as diagnosis and treatment. The concept of AI is relatively easy to grasp. At its core, AI is about hopes that machines can be perfected as to make better decisions than humans. Computers should mimic the human brain, yet the outcome should be near perfect, resulting in better and faster decisions (Mak and Pichika, 2019). AI uses sophisticated computerized calculations and protocols (machine learning 'algorithms'), which have been defined as 'an application of artificial intelligence designed to automatically detect patterns in data without being explicitly programmed' (Watson et al., 2019, p. 1).

Machine learning algorithms have already shown their potential in the medical field for conditions as diverse as diabetes (Gulsham et al., 2016), cancer (Esteva et al., 2017) or pneumonia (Rajpurkar et al., 2017) to mention a few illustrations. Microsoft's InnerEye provides another good example. This tool offers a graphical user interface to algorithms that help radiologists diagnose cancerous tumours and plan surgical interventions (Wiens and Shenoy, 2018). We are moving rapidly from algorithms for clinical decisions that were rule-based and relatively simple, towards comprehensive AI-based machine learning. Simple algorithms of the past would usually rely on less than ten variables. Modern algorithms, on the other hand, offer real-time predictions based on almost unlimited numbers of variables (Parikh et al., 2019). Machine learning's adaptability also sets it apart from traditional software. According to the US Food and Drug Administration (FDA): 'Adaptive artificial intelligence and machine learning technologies differ from other software as a medical device (SaMD) in that they have the potential to adapt and optimize device performance in real-time to continuously improve health care for patients' (US Food and Drug Administration, 2020).

Regulators of medical devices are now granting authorization to pre-clinical and clinical platforms that rely on AI. This, for instance, includes early warning systems that integrates real-time vital sign data to identify hospitalized patients at risk. The use of electronic health records (EHR) is a major step in the development of future algorithms. As stated in the Norwegian Strategy for Personalized Medicine in Healthcare (2017–2021): 'National systems for secure and appropriate storage of large quantities of data must be developed. Standardized systems are required to build up registries and to allow for effective analysis and sharing of data both nationally and internationally. The Electronic Health Record (EHR) is an important ICT-based tool for healthcare providers' (Helsedirektoratet, 2016, p. 8).

There are, however, significant concerns about the ethics (e.g. equal access, data protection) as well as the rigorousness of the methods used to collect and process the information. AI systems, to be safely deployed, must rely on well-understood, realistic and testable assumptions (Balog, 2018).

In view of these possible shortcomings, the question arises as to whether the emergence of algorithms is simply a 'hype' (Parikh et al., 2019) or a revolutionary development that will bear long-lasting fruit. A consensual view, however, is that the success of AI will largely depend on the capacity to integrate genomic risk information in a meaningful way (Mrksic Kovacevic and Bouder, 2020). Or, in other terms, the 'development of appropriate architecture and functionality for handling of genomic data (…) will be pivotal to the successful implementation of personalized medicine' (Helsedirektoratet, 2016, pp. 8–9).

Ramping up access to big data

Progress in AI is likely to raise expectations about the fulfilment of individualized medical treatments, commonly referred to as 'personalized medicine' (Hedgecoe, 2004; Prainsack, 2017). As a move away from a 'one size fits all' approach to medicine, personalized medicine (and the related notions of 'individualized' and 'stratified' medicines) captures the particularities of a patient's disease or predisposition to the disease. Patterns are identified by combining and analysing information about key variables, such as genome and lifestyle. In this context, predictive computerized systems rely on access to large datasets that will then be analysed computationally, i.e. '"Big Data": getting more data, of better quality, from more people (…). Proponents of personalized medicine require access to data from large-scale genomic research projects' (Hoeyer, 2016, p. 532). In the same vein, Adjekum et al. (2017) suggest that: 'To a large extent, precision medicine is driven by three main concurrent trends: the increasing availability of heterogeneous large-scale databases from which novel patient aggregates evolve, advances in the characterization of medically relevant information, and novel computational tools for data analytics' (p. 704).

This is where the medical demands of personalized medicines meet the technical needs of algorithms. To be effective, AI requires access to vast amounts of information derived from robust and reliable sources. A primary enabling component is the availability of large-scale structured knowledge repositories, which organize information around specific objects (Balog, 2018). A crucial question for AI's medical applications is therefore whether the right information is available, both quantitatively and qualitatively, to allow the development of useful algorithmic tools that will support expert judgement as well as patients' needs.

The type of information that may be collected and used for AI modelling potentially includes near infinite amounts of data about the disease, about patients' history, or about patients' genomics. Genomic tests, in particular, have started to play an increasing role in clinical practice. Tests, such as MammaPrint and Onctotype DX, require the collection and processing of considerable amounts of genomic information which is then translated into algorithms (Xin et al., 2017; Mrksic Kovacevic and Bouder, 2020).

This process, however, is not without its hurdles and challenges. AI's incremental nature demands that a constant flow of data be generated to 'feed' and 'perfect' the algorithms. A critical part of AI's added value resides in its perfectibility. The clinical environment (e.g. hospitals) plays a well-known role as a reliable supply source. Large-scale genomic research projects (Hoeyer, 2019) rely on information collected in practices and clinical settings, including, for instance, in the course of routine care (Green et al., 2019). This role was clearly identified from the early days of personalized medicines. For instance, pioneering national programmes such as those of Estonia (1999), and Finland (2006) offered access to clinical data. A considerable amount of clinical data is actively collected, including genetic sequences (Nicholson Price II, 2017).

The regulatory approval process is under greater scrutiny at a time where an increasing number of commercial entities are seeking regulatory approvals for their algorithms (Parikh et al., 2019). In particular, there is a growing interest for the role that policies, and the regulatory environment, play in this context. What is the regulator's role? Is it to enable access to more data or is it to control it? The right answer probably implies it is to fulfil both functions – encouraging the developments that benefit patients while discouraging those that may hinder or mislead them. On the enabling side, regulators have been eager to show that they are on top of technological developments, that they are determined to allow and speed up the supply of data, not only as volumes but also in areas that are particularly needed, as the FDA's Digital Health Innovation Action Plan (2017) illustrates.

One the control side, there is some degree of disagreement about whether the current approach is fit for purpose, whether it is doing too much or too little. One view is that algorithms have not been subjected to the same level of standards, including safety standards, as, for instance, clinical trials (Parikh et al., 2019). This, however, may not apply to all regulators: '(FDA) has suggested that it will regulate algorithms under its traditional framework' (Nicholson Price II, 2017, p. 421). Yet, maintaining a traditional approach raises issues of its own for Big Data. One concern is whether algorithms' pose specific types of issues and should therefore be subjected to different regulatory standards. The current approaches to safety and clinical trials may not fulfil this need, especially in the context of fast-moving technologies and the development of artificial intelligence. In this case, maintaining a 'relatively rigid system' may only 'stifle innovation and to block the development of more flexible, current algorithms' (Nicholson Price II, 2017). A more adaptive regulatory approach may be the way forward, one that is based on transparency. A move in this direction would require developers to disclose information underlying their algorithms. 'Disclosure would allow FDA oversight to be supplemented with evaluation by providers, hospitals, and insurers. This collaborative approach would supplement the agency's review with ongoing real-world feedback from sophisticated market actors' (Nicholson Price II, 2017, p. 421). A well-thought out transparency approach will include ethical considerations, especially as to how far do we take automation and privacy issues, and how to deal with (false) positives and negatives. It will also include the reliability of the information and the communicating of uncertainty.

Transparency and big data: a holy alliance?

Research standards have been established to guide the development and validation of multivariable prediction, for instance the *Transparent reporting of a multivariable prediction model for individual prognosis or diagnosis* (TRIPOD) checklist. Parikh et al. (2019) suggest that, to be acceptable new algorithms would need to meet a series of requirements:

1. Unlock the potential for advance analytics.
2. Protecting patient's safety.
3. Meet standards of clinical benefits – alike predictive biomarkers for instance.

Meeting these expectations will greatly depend on the regulatory infrastructure. There is a need to conduct a discussion about predictive analytics that moves beyond narrowly defined concerns such as tests and advertising (Kalokairiou et al., 2017; Niemec et al., 2017). For instance, the regulation of algorithms includes asking developers to demonstrate safety and efficacy without destroying the flexibility and ongoing innovation that drives medicines' development (Nicholson Price II, 2017). There is a need to pay more attention to external regulatory trends that will affect how Big Data can be made to work at patient level. In this respect, discussions about the implications of the European Union's General Data Protection Regulation (GDPR) for the availability of data is of great interest as it now allows patients to give their explicit consent for use of personal data and as such is forcing a rethink of how we 'do business' in medicines (Haug, 2018).

The move towards 'Regulatory Transparency' is likely to constitute a major development bearing meaningful consequences. Regulatory transparency is a process by which information is made widely available, and as such is an integral part of the 'Big Data' trend. It has been defined as the 'transparency obligations of the state and the principle of fair and equitable treatment' (Kotera, 2008). Although an integral part of the government accountability associated along with accessibility and effective enforcement (Bessette, 2001), it has only become a priority in the past ten years or so. The call for more transparency in the medical field first emerged as a response to a number of questionable practices observed on both sides of the Atlantic (Löfstedt and Bouder, 2014), such as, but not limited to, conflicts of interests, as well as industry concealing the negative results of certain drug trials (IOM, 2006). In the US, criticism led the FDA to actively tackle conflicts of interest and commit to transparency (US Food and Drug Administration, 2007; Hamburg and Scharfstein, 2009; US Food and Drug Administration, 2010). In Europe, regulatory secrecy came under the spotlight in 2007 when the European Medicines Agency (EMA) refused to release data from 15 clinical trials (Gøtzsche and Jørgensen, 2011). In October 2014, after much discussion and concertation, the EMA launched its 'landmark' clinical reports policy (Bouder et al., 2015; Way et al., 2016; Löfstedt and Schlag, 2016). The policy provided 'unprecedented' access to the data and information found in studies investigating the safety and efficacy of medicines (i.e. clinical trial data). This includes:

1 Maintaining publicly accessible repositories and web portals that contain vast amount of patient information, from clinical trials to safety-related data on adverse drug reactions.
2 Publishing agency's supportive information such as recommendations from safety assessments (e.g. periodic safety update reports) lay summaries of risk management plans, and lists of medicinal products being monitored (Löfstedt and Bouder, 2014).

In EMA's own words: 'EMA is committed to continuously extending this approach' (European Medicines Agency, 2014, p. 1). This strongly suggests that the global 'train of transparency' has left the station and is unlikely to return, despite a 'pause' in regulatory transparency initiatives triggered by the UK exit from the European Union ('Brexit').

Why transparency? The EMA's main goal has been to build trust in the agency, including its evaluation system (European Medicines Agency, 2010, 2014; Eichler et al., 2012; Eichler et al., 2013; Koenig et al., 2015; Way et al., 2016): 'the Agency has embarked on this process because it believes that the release of data, making it accessible to all who wish to see it, is about establishing trust and confidence in the system' (European Medicines Agency, 2014, p. 1). Yet, it is also clear that, apart from regulators, the main proponents of this policy have been its direct beneficiaries, especially data miners and academic journals who are willing to conduct and publish reanalyses (Bouder et al., 2015).

Algorithms in the transparency age: tensions

The regulatory transparency policies developed on both side of the Atlantic have resulted in the release of a considerable amount of data derived from clinical trials and other sources, often amounting to several thousands of pages (Way et al., 2016). This material, combined with clinical sources containing crucial genomic information, bears great potential for the development and improvement of future algorithms. Moving forward, building more bridges to link these different sources and cross-using databases in a productive way will be instrumental for making algorithms more accurate. For example, in the context of the ambitious 'All of Us' programme of the National Institute of Health (NIH), hugely different sources are combined, ranging from molecular data to lifestyle data for instance (Anon, 2019).

A qualitative leap in the direction of active communication between experts and non-experts is also necessary to support this trend. As Watson et al. (2019) put it, 'The long-term success [of algorithms] hinges on the ability of both patients and doctors to understand and explain their predictions, especially in complicated cases with major healthcare consequences' (Watson et al., 2019, p. 1). One important concern, however, is that AI has not taken this need very seriously. Back in 2012, Soini noted that the 'asymmetry on information regarding genomic information both among lay people and general practitioners should be given a due regard' (Soini, 2012, p. 152). Too little has been achieved, despite tremendous advancements on the technological side.

A crucial step is to address issues pertaining to the *accessibility* as well as *comprehensibility* of the data. Meeting this objective, however, requires a fundamental re-think, both from the perspective of the methods used to develop algorithms as well as the current approach to regulatory transparency. From an AI perspective, the approach that prevails among the commercial entities expects users to trust systems almost blindly, despite the fact that these systems offer little transparency regarding their inner workings (Balog, 2019). This 'Black-Box' approach (Nicholson Price II, 2017; Watson et al., 2019) creates obvious issues for both accessibility and comprehensibility of the information, as it fundamentally limits the potential for elicitation during doctor–patient communications. Pharmaceutical regulators, on the other hand, have also failed to engage with societal demands for communication. When making information available and transparent, regulators often fail to explain their actions and decisions, including explicitly discarding alternative courses of actions (Coglianese, 2009). In practice, their goal is to achieve 'fishbowl transparency', broadly defined as full disclosure of information, rather than 'reasoned transparency' involving explanatory information and contextualization (Coglianese, 2009; Löfstedt and Bouder, 2014; Bouder et al., 2015). In a recent article, Coglianese and Lehr (2019) showed how government information supporting the machine learning algorithms developed for 'diagnosing medical conditions, operating motor vehicles, and detecting credit card fraud' (p. 3) is essentially of the fishbowl variety. Therefore, unless there is a drastic change of approach, commitments to make algorithms more transparent will be insufficient to address the Black-Box problem. More data might be released, yet algorithms will remain fundamentally opaque and unable to support doctor–patient communication.

The GDPR legislation in Europe, which was primarily designed to protect personal information, might offer an unexpected legal push in the direction of mandating some form of risk communication. To date, organizations and companies facing this new regulation have invested time, effort and money into developing instruments that protect the owners (i.e. patients) without implying their direct involvement. This is typically the case of Blockchain and Cloud storage technologies (Haug, 2018). Yet, the UK's national institute for data science and artificial intelligence, the *Alan Turing Institute*, has also argued that the new legislation is also offering citizens a right to explanation (Alan Turing Institute, 2018). Similar discussions about new procedural obligations and their impact on patients are taking place in the US (for an illuminating discussion see Coglianese and Lehr, 2019). For instance, the bill introduced by Senator Wyden (the *Algorithmic Accountability Act of 2019*) proposes to mandate an assessment of the risks posed by automated decision systems to the privacy or security of personal information. In the past, risk assessment has been a strong driver of cost–benefit analyses as well as risk communication. These new obligations may prove attractive to regulators as they will indirectly help them to fulfil the set goals that disclosure alone is unable to deliver, i.e. increasing trust in their regulatory work as well as empowering patients.

Next steps: involving decision sciences

In a short programmatic article, Mirnezami et al. (2012) urged to introduce new 'training paradigms' for doctors as well as consumer/patient engagement as a way to 'prepare for precision medicine'. Almost a decade later, an increasing number of scholars are warning us against the limitations and dangers of the Black-Box, and yet too little has been achieved. In 2018, a systematic review of the emerging challenges for the regulation of precision medicines concluded that further research is needed to devise an integral approach involving scientific, ethical, legal and social consideration (Chang and Colonna, 2018). In a recent study, we also found that, on the technical side, the developers of algorithms share these concerns: (i) current algorithms are not satisfactory from the perspective of their clinical use and (ii) they inadequately address risk communication among stakeholders (Mrksic Kovacevic and Bouder, 2020). What to do next? Urgent action is needed to better connect the development of medical decision algorithms with decision sciences. Decision sciences, in particular drawing on about 40 years of cognitive research (Fischhoff, 2009; Fischhoff et al., 2011; Fischhoff, 2013),. Particular attention should be paid to:

1 Quality of data: understanding people's perceptions, expectations and preferences, crucially risk perception to check societal acceptability. Use this knowledge to build algorithms that take these preferences on board.
2 Evaluate the impact of algorithms as communication tools. Invest time and effort into explaining what the algorithms mean. Then test their comprehensibility on prescribers and patients.
3 Establish training courses and hold regular conferences that bring together geneticists, regulators, doctors and patients to discuss AI in a more open way. It is important to go beyond narrow circles of expertise.
4 Encourage geneticists and algorithm developers to join the Society for Risk Analysis (SRA), as the key transversal organization with a dedication to bringing together risk scientists (including modellers), Government, Industry and other members of society.

Take home messages:

- Big Data implies access to vast amounts of medical information that may be translated into decision algorithms.
- Current transparency policies are likely to accelerate this trend.
- Yet, algorithms remain opaque and unintelligible to most people (Black-Box) which creates many issues from data protection to comprehensibility.
- Risk and decision sciences support shared outcomes. They offer conceptual frameworks and procedures to connect expert developers, the medical profession and consumers/patients.
- Questions for teaching purposes.
- What role does Big Data play in achieving personalized medicines?
- What is the potential and what are the limitations of medical decision algorithms?
- What are the pros and cons of 'fishbowl transparency'?
- What role can Risk Communication (as a branch of risk science) play for developing and improving current and future algorithms?

References

Adjekum, A., Ienca, M. and Vayene, E. (2017) 'What Is Trust? Ethics and Risk Governance in Precision Medicine and Predictive Analytics', *OMICS*, vol 21, no 12, pp. 704–10.

Alan Turing Institute (2018) 'Impact Story. A right to explanation', www.turing.ac.uk/sites/default/files/2018-07/impact-story-a-right-to-explanation.pdf, accessed 26 April 2020.

Anon (2019) 'The "All of Us" Research Program'. Special Report, *New England Journal of Medicines*, vol 381, no 7, pp. 668–76.

Balog, K. (2018) *Entity-Oriented Search*. Springer Open, Cham.

Bessette, J.M. (2001) 'Accountability: Political', in N.J. Smelser and P.B. Baltes (eds.) *International Encyclopedia of the Social and Behavioural Sciences*. Elsevier, Amsterdam, 38–41.

Bouder, F., Löfstedt, R, Way, D. and Evensen, D. (2015) 'Transparency in Europe: A Quantitative Study', *Risk Analysis*, vol. 35, issue 7, pp. 1210–29.

Chang, L-C. and Colonna, T.E. (2018) 'Recent updates and challenges on the regulation of precision medicine: The United States in perspective', *Regulatory Toxicology and Pharmacology*, vol 96, pp. 41–7.

Coglianese, C. (2009) 'The Transparency President? The Obama Administration and Open Government'. Public Law and Legal Theory Research Paper Series, research paper No. 09–18.

Coglianese, C. and Lehr, D. (2019) 'Transparency and algorithmic governance', *Administrative Law Review*, vol 71, no 1, pp. 1–56.

Eichler, H.-G., Abadie, E., Breckenridge, A., Leufkens, H. and Rasi, G. (2012) 'Open clinical trial data for all? A view from regulators', *PLOS Medicine*, vol. 9, issue 4, doi: 10.1371/journal.pmed.1001202.

Eichler H.-G., Petavy, F., Pignatti, F. and Rasi, G. (2013) 'Access to patient-level trial data—A boon to drug developers', *New England Journal of Medicine*, vol 369, issue 17, pp. 1577–9.

Esteva, A., Kuprel, B., Novoa, R.A., Justin, J., Swetter, S.M., Blau, H.M., and Thrun, S. (2017) 'Dermatologist-level classification of skin cancer with deep neural networks', *Nature*, vol 542, pp. 115–18.

European Medicines Agency (EMA) (2010) *The European Medicines Agency Road Map to 2015: The Agency's contribution to science, medicines, health*. London, EMA.

European Medicines Agency (EMA) (2014) 'European Public Assessment Reports: Background and Context', www.ema.europa.eu/en/medicines/what-we-publish-when/european-public-assessment-reports-background-context, accessed 26 April 2020.

European Medicines Agency (EMA) (2020) 'Clinical Data Publication', www.ema.europa.eu/en/human-regulatory/marketing-authorisation/clinical-data-publication.

Fischhoff, B. (2009) 'Risk Perception and Communication', in , R. Detels, R. Beaglehole, M.A. Lansang and M. Gulliford (eds.) *Oxford Textbook of Public Health*, Fifth Edition, pp. 940–52. Oxford University Press, Oxford.

Fischhoff, B. (2013) 'The sciences of science communication', *Proceedings of the Academies of Sciences of the United States of America*, vol 110, Suppl 3, pp. 14033–39, doi: 10.1073/pnas.1213273110.

Fischhoff, B., Brewer, N.T. and Downs, J.S. (2011) *Communicating Risks and Benefits: An Evidenced-Based User's Guide,* www.fda.gov/downloads/AboutFDA/ReportsManualsForms/Reports/UCM268069.pdf, accessed 26 April 2020.

Gøtzsche P. and Jørgensen, A. (2011) 'Opening up data at the European Medicines Agency', *BMJ*, doi: https://doi.org/10.1136/bmj.d2686.

Green, S., Carusi, A. and Hoeyer, K. (2019) 'Plastic diagnostics: The Remaking of disease and evidence in personalized medicine', *Social Science and Medicine*. doi: 10.1016/j.socscimed.2019.05.023. [Epub ahead of print]

Gulsham, V., Peng, L., Coram, M. Stumpe, M.C., Wu, D., Narayanaswamy, A., Venugopalan, S., Widner, K., Madams, T., Cuadros, J., Kim, R., Raman, R., Nelson, P.C., Mega, J.L. and Webster, D.R. (2016) 'Development and Validation of Deep Learning Algorithm for Detection of Diabetic Retinopathy in Retinal Fundus Photoraphs', *Journal of the American Medical Association*, vol. 316, pp. 2402–10.

Hamburg, M.A. and Sharfstein, J.M. (2009) 'The FDA as a public health agency', *New England Journal of Medicine*, vol. 360, pp. 2493–95.

Haug, C. (2018) 'Turning the Tables – The New European General Data Protection Regulation', *The New England Journal of Medicine*, vol. 379, no 3, pp. 207–9.

Hedgecoe, A. (2004) *The Politics of Personalised Medicine: Pharmacogenetics in the Clinic*. Cambridge University Press, Cambridge.

Helsedirektoratet (2016) 'Summary of the Norwegian Strategy for Personalised Medicine in Healthcare', 2017–2021, www.helsedirektoratet.no/rapporter/strategi-for-persontilpasset-medisin-i-helsetjenesten/Summary%20of%20the%20Norwegian%20Strategy%20for%20Personalised%20Medicine%20in%20Health%20Care.pdf/_/attachment/inline/5a6c511c-b245-4546-8dfa-daa057f275dc:f0a88b9e56ddd d83901639bea4de5c04919bf407/Summary%20of%20the%20Norwegian%20Strategy%20for%20Personalised%20Medicine%20in%20Health%20Care.pdf, accessed 26 April 2020.

Hoeyer, K. (2016) 'Denmark at a crossroad? Intensified data sourcing in a research radical country', in B. Mittelstadt and N. Keiding (eds.) *The Ethics of Biomedical Big Data*. Springer, Dordrecht, 73–93.

Hoeyer, K. (2019) 'Data as promise: Reconfiguring Danish public health through personalized medicine', Social Studies of Science, vol 49, no 4, pp. 531–55

Institute of Medicine (IOM) (2006) *The Future of Drug Safety: Promoting and protecting the health of the public*. National Academies Press, Washington, DC.

Kalokairiou, L., Borry, P. and Howard, H.C. (2017) 'Regulating the advertising of genetic tests in Europe: a balancing act', *Journal of Medical Genetics*, vol. 54, pp. 651–56.

Koenig, F., Slattery, J., Groves, T., Lang, T., Benjamini, Y, Day; S., Bauer, P. and Posch, M. (2015) 'Trends in global clinical trials registration: an analysis of numbers of registered clinical trials in different parts of the world from 2004 to 2013'. *BMJ Open*, vol 5, issue 9, doi: 10.1136/bmjopen-2015-008932.

Kotera, A. (2008) 'Regulatory Transparency', in P. Muchlinski, F. Ortino, and C. Schreuer (eds.) *The Oxford Handbook of International Investment Law*. doi: 10.1093/oxfordhb/9780199231386.013.0016.

Löfstedt, R. and Bouder, F. (2014) 'New Transparency policies: Risk communication's doom?', in J. Arvai and L. Rivers III (eds.) *Effective Risk Communication*. Earthscan from Routledge, Oxon and New York.

Löfstedt, R. and Schlag, A.K. (2016) 'Looking back and going forward: what should the new European Commission do in order to promote evidence-based policy-making?', *Journal of Risk Research*, vol 20, issue 11, pp. 1–20.

Mak, K.K. and Pichika, M.R. (2019) Artificial intelligence in drug development present status and future prospects, *Drug discovery today*, vol 24, no 3, pp. 773–80.

Mirnezami, R., Nicholson, J. and Darzi, A. (2012) 'Preparing for precision medicine', *New England Journal of Medicines*, vol 366, issue 6, pp. 489–91.

Mrksic Kovacevic, S. and Bouder F. (2020) 'Uncertainty in the world of precision medicine: what is an algorithm', *forthcoming*.

Nicholson Price II, W. (2017) 'Regulating Black-Box Medicine', *Michigan Law Review*, vol 116, pp. 421–74.

Niemec, E., Kalokairinou, L. and Howard, H.C. (2017) 'Current ethical and legal issues in health-related direct-to-consumer genetic testing', *Personalised Medicines*, vol 14, no 5, pp. 433–45.

Parikh, R.B., Obermeyer, Z. and Navathe, A.S. (2019) 'Regulation of predictive analytics in medicine', *Science*, vol. 363, issue 6429, pp. 810–12.

Prainsack, B. (2017) *Personalized Medicine. Empowered Patients in the 21st Century*. Cambridge University Press, Cambridge.

Rajpurkar, P., Irvin, J., Zhu, K., Yang, B., Mehta, H., Duan, T., Ding, D., Bagul, A., Ball, R.L., Langlotz, C., Shpanskaya, K., Lungren, M.P. and Ng, A.H. (2017) 'ChXNet: Radiologist-level pneumonia detection on chest X-Rays with deep learning', *arXiv*, 1711.05225, https://arxiv.org/pdf/1711.05225.pdf, accessed 26 April 2020.

Soini, S. (2012) 'Genetic testing legislation in Western Europe- a fluctuating regulatory target', *Journal of Community Genetics*, vol 3, no 2, pp. 143–53.

US Food and Drug Administration (US FDA) (2007) *The Future of Drug Safety-promoting and protecting the health of the public*. FDA's response to the Institute for Medicine's 2006 report. US FDA, Rockville, MD.

US Food and Drug Administration (US FDA) (2010) *FDA Transparency Initiative: Draft proposals for public comment regarding disclosure policies of the U.S. Food and Drug Administration*. US FDA, Rockville, MD.

US Food and Drug Administration (2020) *Artificial Intelligence and Machine Learning in Software as a Medical Device*. Content current as of 28 January 2020. Accessed on 26 April 2020 from: www.fda.gov/medical-devices/software-medical-device-samd/artificial-intelligence-and-machine-learning-software-medical-device

Watson, D.S., Krutzinna, J., Bruce, I.N., Griffiths, C.E.M., McInnes, I.B., Barnes, M.R. and Floridi, L. (2019) 'Clinical applications of machine learning algorithms: beyond the black box', *British Medical Journal*, vol 364, l886. doi: 10.1136/bmj.l886.

Way, D., Bouder, F., Löfstedt, R. and Evensen, D. (2016) 'Medicines transparency at the European Medicines Agency (EMA) in the new information age: the perspectives of patients', *Journal of Risk Research*, vol 19, issue 9, pp. 1185–215.

Wiens, J. and Shenoy, E.S. (2018) 'Machine learning for healthcare: on the verge of a major shift in healthcare epidemiology', *Clinical Infectious Diseases*, vol. 66, pp. 149–53.

Xin, L., Liu, Y.-H., Martin, T.A. and Jiang, W.G. (2017) 'The era of multigene panels comes? The clinical utility of Oncotype DX and Mammaprint', *World Journal of Oncology*, vol 8, issue 2, p. 34.

11 Epilogue

Mats G. Hansson, Ulrik Kihlbom and Silke Schicktanz

Introduction

Since the research program *Mind the Risk* started to take shape several years ago, genomic and biomedical research have developed fast and in ways not very visible at that point of time. Researchers in the fields of next generation genetic sequencing, multi-omic analysis, Big Data, personalized medicine, AI, and epigenetics are now rapidly developing new ways to analyse, diagnose and treat patients in many different disease areas. In contrast, the readiness and ability of ethicists and society at large to respond to the ethical issues in such developments are perhaps less reactive than desirable. It might be suggested that this is what to expect and even argued that this sluggishness of those who's task it is to reflect, is in a way inescapable. This line of argument may refer to Hegel's famous metaphor of how 'the owl of Minerva spreads its wings only with the falling of the dusk'. Hegel formulated it in the preface to his *Philosophy of Rights* and it is often interpreted to mean that our philosophical understanding of scientific and societal processes can only be reached or matured once those developments are in place. As an idea of the precondition of a historical understanding that may well be true. However, scientific as well as political decisions regarding if and how to implement promising biomedical technologies need to and will be made. These are decisions that potentially have a massive impact on the health and well-being of people. Those decisions can be made for better or worse reasons and even if we cannot fully understand and evaluate the current developments, we certainly are obliged to think about and debate possible grounds for such decisions sooner rather than later.

In this last chapter, we will suggest upcoming ethical issues in relation to new developments in biomedicine. We will limit our examples to two cases: next generation sequencing and epigenetics, and point towards ethical issues that will be needed to address in the future. One set of issues concerns how health care professionals and individuals can and should handle increasingly complex and uncertain risk information. Is personal preference and satisfaction sufficient or are clinically actionable routes necessary for performing genomic testing? Can people, including patients as well as health care professionals, in a meaningful way use risk information that is so uncertain that no probabilities can be attributed to it? How are we to think about health care professionals and individuals' responsibilities towards family members when increasingly powerful genomic tests are made available?

Another set of questions is about public health issues around equality and the potential conflict between interests of the individual versus that of the public. Can the implementation of personalized medicine increase unequal access to health care? As we write this epilogue, the virus SARS-CoV-2 is spreading across the world and the pandemic is too clearly throwing the ethically contested nature of public health interventions into sharp relief. What limitations of individual freedom can be justified for the sake of safeguarding the health of the general public or some high-risk groups? How are to think about individual responsibilities regarding the health of coming generations?

These questions are hard to answer in ways that find general agreement. The difficulty of formulating substantial resolutions to ethical problems on a societal level has led many philosophers towards a procedural approach. Much of that started with John Rawls' *Theory of Justice* (Rawls 1971), and in health care ethics, the work of Norman Daniels and James Sabin's work has been important (Daniels and Sabin 1997). On the societal level it might be easier to reach agreement on a fair and ethically acceptable process for how to arrive at decisions on ethically contested issues than on substantial principles. One process-oriented way to respond to challenging ethical issues might be public and patient involvement. It is not something that can be considered as *the* answer but might be part of *an* answer. Public and patient engagement has grown considerably in recent years and continues to do so. But however ethically promising this development seems to be, it will also need careful and critical scrutiny. We conclude this chapter with some reflections on how public and patient involvement may be critically assessed.

Challenges when Next Generation Sequencing (NGS) is introduced in routine clinical care

Traditionally, genetic testing was confined to specialist medical genetic services, focused on relatively rare, high penetrance inherited diseases. In contrast, the common, complex disorders such as heart disease are usually the result of variation in many genes, each contributing a small amount of genetic susceptibility, acting in concert with environmental or epigenetic factors. A recent development in high-throughput genetic health care technologies is capable of generating large volumes of genetic risk information, including information about unsolicited findings. Whilst this development gives rise to hopes of individualized health advice and selection of optimal treatment and prevention, the information will be more complex as well as having less predictive power. Moreover, diagnosing people with a genetic disease may give rise to a range of ethical issues (Chapter 7 in this volume; Silverstein et al. 2014). Understanding and dealing with genetic information is influenced by cultural and educational differences, and the public in general have limited understanding of genetic information (Chapters 8 and 9 in this volume; Condit 2010) which makes clinical genetics a further challenge and emphasizes the need of ensuring patient benefit. Following this rapid progress in genome sequencing, genetic information will, in an increasing degree, be relevant in clinical settings and provide more precise and personalized diagnoses and treatments for patients. However, with this progress comes the obligation to ensure that providing patients with genetic risk information leads to patient benefit. In complex interventions such as clinical genetic services there are other relevant outcomes to consider besides health gain, especially for genetic diagnoses where no treatment is available. An unforeseen and often neglected consequence of personalized medicine is that patients, with the same condition but with different genetic compositions, can be offered different treatment regimes. This inequality in access to treatment gives rise to concerns of fairness, which, from both a health economic and ethical perspective, is highly relevant and needs to be considered by society and in priority settings. We will give two examples of introducing NGS in clinical settings, from the area of cardiogenetics and from oncogenetics.

Monogenic cardiomyopathies associated with risk of sudden death have over the recent years become an emerging field that is rapidly developing, including cardiomyopathies with structurally detectable substrate (hypertrophic cardiomyopathy or arrhythmogenic right ventricular cardiomyopathy) or those with substrate confined to an ultrastructural level of ion channels, so called channelopathies (such as long or short QT syndrome [LQTS], Brugada syndrome, cathecholaminergic polymorphic ventricular tachycardia [CPVT]). The common feature for all these conditions is their relatively high prevalence in the population (from 1:2000 for LQTS to 1:5000 for Brugada syndrome), the autosomal dominant inheritance pattern, and the risk of sudden arrhythmic death, which may be the first symptom occurring before the age of 40. It is estimated that 2–3 of 1000 in the population have a phenotype of one of the most common genetically determined cardiomyopathies. However, the vast majority of affected individuals have a mild phenotype and are asymptomatic, which further complicates early detection of the disease and proper screening for risk individuals in affected families, as well as the decision to go through with carrier testing (pre-symptomatic genetic testing). Genetic counselling and pre-symptomatic genetic testing of first-degree relatives have become an essential tool for early identification of individuals at risk and are now offered in specialized cardiogenetic settings. The effects of providing genetic risk information have traditionally been measured with the help of various kinds of Quality of Life instruments (Kettis Lindblad et al. 2002). However, these are too blunt for capturing what may be at stake in association with the provision of genetic risk information where there is sometimes nothing or very little that is clinically actionable.

Studies from large cohorts of cancer patients show that approximately 8–10 per cent of all adult cancers are hereditary (Huang et al. 2018). Today there are more than 150 known genes that cause primary cancer syndromes and the list gets even longer if benign tumours and haematologic malignancies are included. Massive parallel sequencing (MPS), a DNA sequencing method that enables sequence analysis of the entire exome or genome, has been of great help to discover new high-risk genes for cancer (Tesi et al. 2017; Adam et al. 2016).

In haematology, several studies have shown that hereditary haematological malignancies, e.g. AML/MDS are much more common than previously thought (Zhang et al. 2015; Wartiovaara-Kautto et al. 2018), which led to the inclusion of hereditary haematologic malignancies as a subgroup in WHO classification (Arber et al. 2016) and several genetic changes have been identified as the cause. In Sweden, at least 50 families have been identified over the last 2–3 years. This number is expected to increase as more patients are investigated for these diseases. There is probably a large proportion of unrecorded cases because the awareness of hereditary predisposition to haematological malignancies is low. Knowledge of the usefulness of identifying families, in the form of improved prognosis or increased sense of security, is thus also inadequate.

Another unusual hereditary cancer syndrome is Li Fraumeni Syndrome (LFS), caused by mutations in the gene *TP53* (McBride et al. 2014). The syndrome is characterized by family trees with many different types of tumours, including sarcoma, leukaemia in children, very early breast cancer and brain tumours, and it has been difficult to have an adequate follow-up program with the exception of breast cancer surveillance. Therefore, pre-symptomatic testing was not always offered, especially to men and children because it was considered that it would only cause concern. However, recent studies have shown that follow-up improves prognosis in these families (Kratz et al. 2017). Moreover, approximately one third of the families with germline *TP53* mutations tend to only develop breast cancer in females, and apparently not the other tumours associated with LFS. Therefore, it has been suggested to refer to this syndrome as *heritable TP53-related cancer syndromes* (Frebourg et al. in press). *TP53* involves families of index patients to a hitherto unexperienced extent, putting extra pressure on clinicians to provide counselling, not only to the index patient but to whole families.

Another new group of persons currently being analysed for genes responsible for hereditary cancer are women who have breast and/or ovarian cancer where new drugs such as ADP-ribose polymerase inhibitors (PARPP inhibitors) are currently becoming available (Taylor et al. 2018; Tung and Garber 2018). These are effective only if gene changes exist in the *BRCA* genes in the tumour itself. Some of these changes are somatic while others will prove to be constitutional, i.e. hereditary. Depending on the genetic profiles of women with a diagnosed breast cancer a relatively substantial number of women are relevant for this treatment-guided testing. Triple negative Breast Cancer (TNBC) is one such group accounting for 9 per cent of all new BCs. With approximately 8000 new Swedish cases per year, 720 of these will be TNBC. Twenty per cent of these have loss of function mutations in *BRCA1/2*, implying 144 women are affected each year. Added to this are other subtypes, e.g. hormone receptor-positive HER2-negative BC. *BRCA1/2* mutation testing may help to identify patients with BC who might benefit from treatment with PARPP. However, selection of treatment is complex due to differences in prospective progression free survival, overall survival and adverse events related to the treatment. For some mutations it is not yet clear whether the new treatments are superior to conventional anthracycline-based or taxane-based chemotherapy (Tung and Garber 2018).

Currently clinical geneticists are receiving an increasing number of referrals from oncologists and haematologists who expect or suspect a hereditary profile of a patient's cancer. A special challenge regarding this development is related to the need to manage risk information with smaller or greater degrees of uncertainty attached. The goal is to target treatment to individual patients' needs but in many instances there is yet not much to offer, besides information and monitoring, hoping that this will at least provide a sense of control and in some instances lead to better prevention strategies.

Challenges ahead in the wake of epigenetics

There is growing evidence of the impact of non-genetic variables on gene expression, e.g. nutrition, maternal care/behaviour, psychosocial stress, adversity and neglect in early life, hormones and drugs. Epigenetic mechanisms (DNA-methylation, histone modification and aberrant use of micro-RNA) explain how the exposure of environmental factors influence the phenotypic outcome and variability both between individuals and within an individual at different times. As suggested by Frances Champagne the dynamics of epigenetics is likely of great evolutionary significance. 'The plasticity of DNA methylation and posttranslational histone modifications in response to both postnatal and adult social experiences suggest that these mechanisms may have evolved to allow organisms to adapt to changing environmental conditions' (Champagne 2012).

Nutrition

Barker explained the concept of foetal growth affecting adult disease in two seminal papers (Barker 1992; Barker 1995). Numerous studies have since then been carried out in order to investigate the effects of early dietary exposure on adult health. Health records collected during the Dutch Hunger Winter of 1944–1945 allowed analysis of long-term consequences of prenatal famine (Stuffrein-Roberts et al. 2008). An association was found between severe nutritional deficiency and schizophrenia (Susser et al. 1996). The relationship was replicated by analysis of health records of the famine in China during 1959–1961 (St Clair et al. 2005). There remain considerable gaps in the knowledge of causal relationships between perinatal nutrition and health endpoints at later stages in life through epigenetic mechanisms, and there may be many confounders that must be identified and controlled for, but evidence on metabolic imprinting/programming is building up rapidly (Hanley et al. 2010). Recent studies point also to a relationship between early life precursors, epigenetics and food allergy (Hong and Wang 2012).

Adversity, neglect in early life and social environments

That epigenetic programming may be caused by maternal behaviour has been demonstrated in rat models. Increased pup licking, grooming and arched-back nursing by rat mothers altered the offspring epigenome, affecting differences in stress reactivity across generations (Weaver et al. 2004). The effects were reversed through cross fostering indicating reversibility in programming through changes in behaviour. Chris Murgatroyd and Dietmar Spengler demonstrated recently that adversity in early life might shape the experience-dependent maturation of stress-regulating pathways underlying emotional functions (Murgatroyd and Spengler 2011). Social experience-dependent DNA memory like this is suggested to follow a sequential action from an epigenetic cue emanating from the social and/or physical environment, through intracellular pathways to chromatin/DNA response (Hoffmann and Spengler 2014).

As described by Frances Champagne, decades of longitudinal and laboratory-based studies have highlighted the relationship between the social environment and behavioural and health outcomes (op.cit.) There are several epidemiological registry-based studies pointing in the same direction (Modin et al. 2008). Continuous results in this direction are coming from the Stockholm Public Health Cohorts as well as other registries (Svensson et al. 2012).

Psychiatric disorders

There are several examples of environmental factors, both during prenatal and childhood/adolescence developmental stages, that have been proposed to interact with genetic factors in major psychotic disorders (Rutten and Mill 2009). As noticed by Bart Rutten and Jonathan Mill (ibid.), however, they often represent only statistical interactions and there is a lack of knowledge about etiological mechanisms even if several epigenetic mediators have been found in animal research. Psychotic syndromes may be understood as disorders of adaption to a social context, implying that one needs to understand both the social context and the cellular and genetic factors. A lesson from epigenetics is that heritability estimates from classical twin studies reflect both the genetic influence and the underlying gene-environment interactions (Van Os et al. 2010). As concluded by Frances Champagne, 'epigenetic mechanisms play a critical role in development and may serve both to shape development in response to social experiences and to induce variation in social behaviour' (*op.cit.)*.

Potential for social improvement and pharmacological treatment

Epigenetics opens the way for understanding and improving the social environment but provides also new pathways and targets for pharmacological treatment. Unlike genetic mutations, if a phenotype is caused by epigenetic phenomena, such as DNA methylation and histone acetylation, they can be chemically reversed (Seo et al. 2013). The advantage of developing and bringing epigenetic drugs into therapeutics is the reversible nature of epigenetic pathways (Boks et al. 2012). The same holds for social improvement and treatments directed to change behaviour.

The case of obesity

The prevalence statistics on childhood obesity have escalated rapidly during the last two decades in almost all countries irrespective of their economic status (WHO 2014). It is argued that country-specific trends are now coalescing to create a true pandemic, penetrating also the poorest nations in the world (Prentice and Jebb 2006). Changing individual lifestyle is generally regarded as a primary measure for prevention of obesity and several other non-communicable diseases (Lee 2016) but the efficacy of current approaches is limited. Yet, there is growing (for a historical review see Gluckman et al. 2010; Wadhwa et al. 2009) and compelling evidence of early developmental (from preconception through infancy) influences on the risk of the offspring subsequently developing obesity and its associated morbidities (Gluckman et al. 2007; Gluckman and Hanson 2004). Indeed, a number of direct and indirect mechanisms can serve to transmit the effect of environmental influences from one generation to another and are generally termed parental effects (Gluckman et al. 2010). Studies using historical records have shown that food supply in one generation will affect health for at least the subsequent two generations (Susser et al. 1996; Stuffrein-Roberts et al. 2008; St Clair et al. 2005). The effects of food supply on offspring and grandchild mortality risk ratios was analysed by using 303 probands and their 1818 parents and grandparents from the 1890, 1905 and 1920 Överkalix cohorts (Pembrey et al. 2006). Paternal grandfather's food supply was associated with mortality risk ratios of grandsons while paternal grandmother's food supply was associated with the mortality risk ratios of granddaughters.

There is now compelling animal and growing human evidence that at least some of these trans-generational and parental effects are mediated by epigenetic mechanisms (Gluckman and Hanson 2004). Of particular note, epigenetic changes in specific sites in the promotor of metabolism-related genes, such as RXRa in DNA collected from the infant at birth, strongly correlate both with maternal nutritional state in early pregnancy and with offspring adiposity at 6–9 years of age (Godfrey et al. 2011). Intergenerational epigenetic effects induced by developmental challenges are well reported in rodents and are supported by molecular studies (Painter et al. 2005) in second-generation offspring of Dutch famine survivors and clinical observations (Mathai et al. 2013) in offspring of programmed children of both sexes where programming was induced by prematurity. While most research has focused on maternal effects, both experimental studies (e.g. Ferguson-Smith and Patti 2011) and clinical and epidemiological studies (e.g. Curley 2011; Mathai et al. 2013) provide evidence of phenotypic and epigenetic effects mediated via the paternal line. At the other end of the nutritional spectrum, it is also clear that maternal obesity and gestational diabetes also drive offspring obesity and such influence can also be mimicked experimentally and, again, some of these effects may be mediated epigenetically, while others likely reflect the adipogenic effect of elevated foetal insulin (Gluckman and Hanson 2004).

Genetic and epigenetic mechanisms may account for an uncertain but significant proportion of the aetiology of obesity and type 2 diabetes, and recent estimates drawn from the studies of Godfrey et al. suggest that it may be much higher than generally supposed (Gluckman et al. 2011). This conclusion is further supported by the Working Group Report on Science and Evidence to the WHO Commission on ending childhood obesity (WHO 2014) which has concluded that a life course approach starting before conception will be essential to reducing the growing epidemic of childhood obesity which in turn begets adult obesity. The dominant theoretical paradigm suggests that developmental influences alter the sensitivity of the individual to latter obesogenic exposures – evolutionary explanations have been proffered for why these mechanisms persist (see Hanson and Gluckman 2014).

Social practices, local cultures and family patterns of lifestyle clearly also play a role in both creating an obesogenic environment and in creating a milieu for the developing conceptus or infant that can induce parental effects which may or may not involve epigenetic processes but nevertheless have clear transgenerational effects. When non-contagious diseases become 'contagious' to future generations, the distinction between communicable and non-communicable diseases in the area of public health may become less relevant from an ethical point of view. The balance between individual liberties and health conditions for large population groups may change and lead to arguments for strengthening the social context, especially when the health effects would be positive for both current and future generations. Assuming increasing support for the inducement of parental effects related to obesity in response to developmental environmental exposures and lifestyle there are significant challenges ahead regarding public health and ethics.

Epigenetics provides a renewed focus on intergenerational relationships since not only genetics but also lifestyle factors may affect the offspring in more than one generation. Mothers may be alerted on the significance of lifestyle and diet during pregnancy but not before becoming pregnant. That fatherhood may also demand attention to lifestyle and diet is not a widely accepted idea. We tend to start thinking about parental responsibilities when the child is conceived, and about grandparental responsibilities when the first grandchild is born. From an epigenetic point of view, this may be too late. Different cultures may have different views on parents and grandparents as responsible parties to the well-being of children, and to the significance of the family's genetic and social lineage. There is a social construction of parenthood and childhood which will define what constitutes a 'good mother', 'good father', 'healthy child'.

One may ask if epigenetic information will change the views on individual responsibility regarding lifestyle and if society will have new and more extensive responsibilities to intervene in order to modify conceptions of parental responsibility, or family lifestyle factors (e.g. food intake and physical activity) if these can be shown to have epigenetic effects on future offspring. A change in our understanding of the aetiology of obesity could change the ways in which responsibilities for avoiding and treating this condition are attributed among different actors. The 2011 meeting of the United Nations General Assembly and the 2013 WHO Action Plan for the prevention and control of Non-Communicable Diseases adopted a life-course approach that places emphasis on maternal health including in the pre-, ante- and post-natal stages of pregnancy, breastfeeding, and the promotion of health and development for children and adolescents. In addition, responses must be considered to address cumulative risks from transgenerational carry-over effects, as understanding in this field evolves, including whether and how household compositions and family structures may be relevant risk factors for childhood obesity.

Current models for the ethical evaluation of public health measures in modernised health care systems of industrialized countries, such as liberal, communitarian or stewardship frameworks, are not well developed to address the ethics of involving families in health care and social care. In part, this is due to absence of a focus within some of these approaches on parental roles and responsibilities that can inform the ethics of involving families in public health interventions. Liberal arguments against restricting freedom in this way may be weakened if the decisions that individuals make have epigenetic effects that challenge the health of the next generation. It has been suggested that obesity is such a global challenge that one needs to investigate different ways to exert social pressure, e.g. nudging, without stigmatizing individuals (Callahan 2013). How this delicate balance between individual and societal responsibilities is to be achieved is, however, far from evident. Moreover, issues around our responsibilities for future generations raise questions concerning how any burdens for future health gains should be allocated between generations.

Taking public and patient engagement seriously in genomic research and in societal debates

The interest regarding 'Public and Patient Engagement' in the context of data-intensive research in medicine and health care has grown considerably in recent years, especially in regard to the combination of genomic and other health related data booms (Woolley et al. 2016). In general, the respective research is often discussed under the label of 'Big Data'. Although the concept of 'Big Data' is still vague and contested, it usually refers to large data volumes characterized by a high velocity, complexity and variety, whose collection, storage, management and analysis demand new information technological solutions (Gandomi and Haider 2015). In the medical context, this implies that data from clinical routine, research and patient-reported data, but also non-medical, even unstructured, social or demographic data, are aggregated and linked in order to optimize biomedical research. This development triggers high hopes for more personalized treatments, as the merging of large data sets from different sources is expected to facilitate disease prediction as well as treatment decisions through the stratification of patients according to specific disease subgroups. To understand complex diseases, as well as multifactorial interactions between different and potentially disease-related variables, large data volumes must be analysed. This is particularly relevant for rare diseases or diseases with unknown causes. In these contexts, data-driven approaches are deemed more promising than hypothesis-driven research strategies.

The development of public and patient engagement promises a new role for subjects: being approached as partners in research or citizen researchers. However, the meaning and practical implications of such engagement are far from clear. Participation has been characterized as a vague concept (Kelty et al. 2015) and the plurality of participatory terms used in the context of biomedical research exemplifies this confusion, as Beier, Schweda and Schicktanz (2019) argue. That analysis suggests staying with a particular *normative understanding of participation*. Relying on the meaning of participation in political theory and bioethics, three *paradigmatic types of practical conditions* that constitute and define participation were suggested: (a) consent to individual provision of data or biomaterial, (b) consultation regarding research and health policy aims and (c) cooperation in decision-making processes or even as co-researchers.

While public or citizen participation can take on various practical forms, there is a broad consensus that it needs to involve some transfer of power, so that those who are affected by a decision *have an input* into the respective decision-making process. Underlying this is the original political ideal of autonomy in the sense of participation in collective self-government. Thus, political participation should not be equated with mere communication strategies (e.g., the state informs citizens about political strategies or aims), but must grant citizens some real influence which is often understood as empowerment. Further ethical reasons for the inclusion of lay perspectives and lay moralities in social debates over life sciences and medicine, as in the context of empirically informed bioethics, to ensure that the epistemic and experiential perspective of patients and other affected persons are considered adequately.

However, frequent appeals to engagement and participation in the big data and genomic research might sometimes oversee these normative foundations of participation (Blasimme and Vayena 2016). The (positive) future of bioethics and good governance lies in recalling and implementing the different levels of participation from a normative point of view.

To do so, it is important to avoid some pitfalls and shortcomings in using public participation or patient engagement in a purely rhetoric way. If the first level of participation should be still fulfilled, patients now themselves collecting health-related data by using wearables, sensors or health-apps or by simply filling out a questionnaire on a tablet computer does not change that they should have a right to individual consent. Initiatives that in an obscure or even involuntary way collect data cannot claim to let patients 'participate' in their research, they otherwise just reduce patients to bio-data or bio-material collectors. The other end of the spectrum of normative participation would require the involvement of lay participants throughout all phases of the research process. This is still a rarity in big data and genomic research, with an exception such as the US-based private company, *Patients.Like.Me*. This platform asks for patients' data but allows them to compile own research cohorts (Wicks et al. 2011). The empowering co-investigation practices the original political-ethical spirit of participation, namely: the transfer of power. However, such data cooperatives place high demands on participants, as it requires abilities, e.g., social and educational, and resources, e.g. time (Vayena and Blasimme 2017). The promotion of digital health literacy in the broader public may be one step to achieve this (Samerski et al. 2018).

To summarize, and in line with other suggestions (e.g. Beier et al. 2019), the following points will be important to consider to enhance participation and engagement of patients and the public in the future of multi-sources and big data research:

a Projects should critically reflect the underlying motivation for appealing to participatory notions. In particular, it is necessary to distinguish between merely instrumental uses, for example in order to increase the volume of data or acquire financial gains, and genuinely normative reasons for involving patients in order to adopt research aims to patients' opinions. Purely 'token public involvement' (Florin and Dixon 2004) may generate significant mistrust if the public notes failures in achieving this goal. Moreover, it is important to critically reflect on the ethical need of patients or subjects becoming involved as co-researchers. Given the challenges of data privacy or potential misuse in data-intensive research, much would be gained if patients were involved in the *governance* of such research initiatives (e.g. as members of a project advisory board, or regularly invited to workshops).

b It is important to become more transparent about whether patients participate for their own medical benefit or whether participatory notions are evoked in the sense of public engagement that goes beyond individual purposes (Warsh 2014). Often, data-intensive research in medicine and health care are increasingly blurring the boundaries of clinical care and research. Consequently, it is not only important to avoid therapeutic misconceptions, but also participatory misconceptions. Such misconceptions occur if research initiatives use public engagement rhetoric without actually empowering patients as genuine partners in research. In fact, although it may be beneficial to individual patients to receive information on their individual health status, it is misleading to frame this in terms of patient involvement in research.

c One should resist the tendency to depoliticize the notion of participation by denying patients real influence on research – or even exploiting their engagement in order to outsource research work and save time and money (Prainsack 2017). Ethical problems may arise if patients are led to believe that they can take their medical destinies into their own hands, while it is actually others who are in charge.

d If patients are expected to participate in the governance of data-intensive research, research projects should be prepared to hold sufficient financial and structural resources to allow patients, lay persons or patient representatives to effectively communicate and express their opinions. Ultimately, equal access to funding and relevant sources of knowledge production would be a necessary precondition for true power balance and equal status.

e A remaining concern is that even the most sophisticated participatory approaches have a certain affirmative tendency, i.e. an inclination to presuppose and reinforce basic agreement with and support for new technological developments rather than objection or rejection. Hence, on a collective level, the public expression of basic concerns about, or even disagreement with, such technological developments may only be possible outside of the respective initiatives. Therefore societal debates beyond the science–society interaction can be seen still as an important part of modern, democratic societies.

References

Adam, R., Spier, I., Zhao, B., Kloth, M., Marquez, J., Hinrichsen, I., Kirfel, J. et al. 2016. Exome Sequencing Identifies Biallelic MSH3 Germline Mutations as a Recessive Subtype of Colorectal Adenomatous Polyposis. *The American Journal of Human Genetics*, 99(2):337–51.

Arber, D.A., Orazi, A. Hasserjian, R., Thiele, J., Borowitz, M.J., Le Beau, M.M., Bloomfield, C.D., Cazzola, M. and Vardiman, J.W. 2016. The 2016 Revision to the World Health Organization Classification of Myeloid Neoplasms and Acute Leukemia. *Blood*, 127(20):2391–405.

Barker, D.J. 1992. The effect of nutrition of the fetus and neonate on cardiovascular disease in adult life, *Proceedings of the Nutrition Society* 51:135–144.

Barker, D.J. 1995. The fetal and infant origins of disease, *European Journal of Clinical Investigation*, 25:457–63.

Beier, K., Schweda, M. and Schicktanz, S. 2019. Taking patient involvement seriously: critical ethical analysis of participatory approaches in data-intensive medical research, *BMC Med Inform Decis Mak*, 19, 90. https://doi.org/10.1186/s12911-019-0799-7

Blasimme, A., Vayena, E. 2016. Becoming partners, retaining autonomy: ethical considerations on the development of precision medicine, *BMC Med Ethics*, 17:67.

Boks, M.P., de Jong, N.M., Kas, M.J.H., Vinkers, C.H., Fernandes, C., Kahn, R.S., Mill, J. and Ophoff, R.A. 2012. Current status and future prospects for epigenetic psychopharmacology, *Epigenetics*, 7(1):20–8.

Callahan D. 2013. Children, stigma, and obesity, *JAMA Pediatr*, 167(9):791–2.

Champagne, F.A. 2012. Interplay between social experiences and the genome: Epigenetic consequences for behavior, *Advances in Genetics*, 77:33–57.

Condit, C.M. 2010. Public understandings of genetics and health, *Clinical genetics*, 77(1), 1–9.

Curley, J.P., Mashoodh, R. and Champagne F.A. 2011. Epigenetics and the origins of paternal effects, *Horm Behav*, 59(3):306–14.

Daniels, N. and Sabin, J. 1997. Limits to Health Care: Fair Procedures, Democratic Deliberation, and the Legitimacy Problem for Insurers, *Philosophy and Public Affairs*, 4:303–50.

Ferguson-Smith, A.C. and Patti, M.E. 2011. You are what your dad ate, *Cell Metabol*, 13(2):115–17.

Florin, D. and Dixon, J. 2004. Public Involvement in Health Care. *BMJ*, 328.

Frebourg, T., Bajalica-Lagercrantz, S., Oliveira, C., Magenheim, R. and Evans, G. 2020. The European Reference Network GENTURIS. Guidelines for the Li-Fraumeni and Heritable TP53-Related Cancer syndromes, *EJHG, in press.*

Gandomi, A. and Haider, M. 2015. Beyond the hype: Big data concepts, methods, and analytics, *International Journal of Information Management*, 35(2):137–44.

Gluckman, P., et al. 2010. A conceptual framework for the developmental origins of health and disease, *Journal of Developmental Origins of Health and Disease*, 1(1):6–18.

Gluckman, P.D. and Hanson, M.A. 2004. Living with the past: evolution, development and patterns of disease, *Science*, 305(5691):1733–6.

Gluckman, P.D., Hanson, M.A. and Beedle, A.S. 2007. Non-genomic transgenerational inheritance of disease risk, *Bioe Essays*, 29(2):145–54.

Gluckman, P.D., Hanson, M.A., Zimmet, P. and Forrester, T. 2011, Losing the war against obesity: The need for a developmental perspective, *Transl Med*, 3(93):93cm19.

Godfrey, K.M., Sheppard, A., Gluckman, P.D., Lillyprop, K.A., Burdge, G.C., McLean, C., Rodford, J. et.al. 2011. Epigenetic gene promoter methylation at birth is associated with child's later adiposity, *Diabetes*, May;60(5): 1528–34.

Hanley, B., Dijane J, Fewtrell M, Grynberg A. 2010. Metabolic programming, imprinting and epigenetics – a review of present priorities and future opportunities, *British Journal of Nutrition*, 104 Suppl 1:S1–25.

Hanson, M.A. and Gluckman, P.D. 2014. Early developmental conditioning of later health and disease: physiology or pathophysiology? *Physiol Rev*, 94(4):1027–76. doi: 10.1152/physrev.00029.2013

Hoffmann, A. and Spengler, D. 2014. DNA memories of early social life, *Neuroscience*, 264, 64–75. doi: 10.1016

Hong, X. and Wang, X. 2012. Early life precursors, epigenetics, and the development of food allergy, *Seminars in Immunopathology*, 34:655–69.

Huang, Kuan-lin, Mashl, R.J., Wu, Y., Ritter, D.I., Wang, J., Oh, C., Paczkowska, P., et al. 2018. Pathogenic Germline Variants in 10,389 Adult Cancers. *Cell*, 173(2):355–70.e14.

Kelty, C., Panofsky, A., Currie, M., Crooks, R., Erickson, S., Garcia, P., Wartenbe, M. and Wood, S. 2015. Seven dimensions of contemporary participation disentangled, *J Assoc Inf Sci Technol*, 66:474–88.

Kettis Lindblad, Å, Ring, L., Glimelius, B. and Hansson, M. 2002. Focus on the individual. Quality of life assessments in oncology validity, interpretation and implementation in clinical practice (review), *Acta Oncologica*, 41:507–16.

Kratz, C.P., Achatz Maria, I., Brugieres, L., Frebourg, T., Garber, J.E., et al. 2017. Cancer Screening Recommendations for Individuals with Li-Fraumeni Syndrome, *Clin Cancer Res*, 23(11):e38–e45.

Lee, H.J., Klee, S.Y. and Park, E.C. 2016. Do family meals affect childhood overweight or obesity?: nationwide survey 2008–2012, *Pediatr Obes*, Jun;11(3): 161–5.

Mathai, S., Derraik, J.G.B., Cutfield, W.S., et al. 2013. Increased adiposity in adults born preterm and their children, *PLoS One*, 8(11):Nr. e81840.

McBride, K.A., Ballinger, M.L., Killick, E., Kirk, J., Tattersall, M.H.N., Eeles, R.A., Thomas, D.M. and Mitchell, G. 2014. Li-Fraumeni syndrome: cancer risk assessment and clinical management, *Nat Rev Clin Oncol*, 11(5):260–71.

Modin, B., Vågerö, D., Hallqvist, J. and Koupil, I. 2008. The contribution of parental and grandparental childhood social disadvantage to circulatory disease diagnosis in young Swedish men, *Social Science and Medicine*, 66:822–34.

Murgatroyd, C. and Spengler, D. 2011. Epigenetics of early child development, *Frontiers in Psychiatry*, 2:Article 16:1–15.

Painter, R.C., Roseboom, T.J. and Bleker, O.P. 2005. Prenatal exposure to the Dutch famine and disease in later life: an overview, *Reprod Toxicol*, 20(3):345–52.

Pembrey, M.E., Bygren, L.O., et al. 2006. Sex-specific, male-line transgenerational responses in humans, *Eur J Hum Genet*, 14(2):159–66.

Prainsack, B. 2017. *Personalized Medicine. Empowered patients in the 21st century?* New York, New York University Press.

Prentice, A. and Jebb, S. 2006. TV and inactivity are separate contributors to metabolic risk factors in children, *PLoS Med*, 3(12):e481.

Rawls, J. 1971. *A Theory of Justice,* Cambridge, MA, Harvard University Press. Revised edition, 1999.

Rutten, B.P.F. and Mill, J. 2009. Epigenetic mediation of environmental influences in major psychotic disorders, *Achizophrenia Bulletin*, 35(6):1045–56.

Samerski, S., Hofreuter-Gätgens, K. and Müller, H. 2018. Refining and promoting digital health literacy for patients and organizations: The 'TK-DiSK' study, *Eur J Public Health*, 28(suppl 4):cky213.648, https://doi.org/10.1093/eurpub/cky213.648

Seo, S. et al. 2013. Epigenetics: A promising paradigm for better understanding and managing pain, *The Journal of Pain*, 14(6):549–57.

Silverstein, L.B., Stolerman, M., Hidayatallah, N., McDonald, T., Walsh, C., Paljevic, E. et al. 2014. Translating advances in cardiogenetics into effective clinical practice, *Qual Health Res*, 24(10):1315–28.

Soh, S.E., et al., 2014. Cohort profile: Growing up in Singapore Towards healthy Outcomes (GUSTO) birth cohort study, *Int J Epidemiol*, 43(5):1401–9

St Clair, D., Xu, M., Wang, P. Yu, Y., Fang, Y., Zhang, F., Zheng, X., Gu, N., Feng, G., Sham, P. and He, L. 2005. Rates of adult schizophrenia following prenatal exposure to the Chinese famine of 1959–1961, *JAMA*, 294:557–62.

Stuffrein-Roberts, S., Joyce, P.R. and Kennedy, M.A. 2008. Role of epigenetics in mental disorders, *Australian and New Zealand Journal of Psychiatry*, 42:97–107.

Susser, E., Neugebauer, R., Hoek, H.W. Brown, A.S., Lin, S., Labovitz, D. and Gorman, J.M. 1996. Schizophrenia after prenatal famine. Further evidence, *Archives of General Psychiatry*, 53:25–31.

Svensson, A.C., Fredlund, P., Laflamme, L., Alfredsson, L., Ekbom, A., Feychting, M., Forsberg, B., Pedersen, N., Vågerö, D. and Magnusson C. 2012. Cohort profile: The Stockholm Public Health Cohort, *International Journal of Epidemiology*, 42(5):1263–72.

Taylor, K.N. and Eskander Ramez, N. 2018. PARP Inhibitors in Epithelial Ovarian Cancer. *Recent Pat Anticancer Drug Discov*, 13(2):145–58.

Tesi, B., Davidsson, J., Voss, M., Rahikkala, E., Holmes, T.D., Chiang, S.C.C., Komulainen-Ebrahim, J., Gorcenco, S., et al. 2017. Gain-of-Function SAMD9L Mutations Cause a Syndrome of Cytopenia, Immunodeficiency, MDS, and Neurological Symptoms, *Blood*, 129(16):2266–79.

Tung, N.M. and Garber, J.E. 2018. BRCA1/2 testing: therapeutic implications for breast cancer management, *Br J Cancer*, 119(2):141–52.

Van Os, J., Kenis, G. and Rutten, B.P.F. 2010. The environment and Schizophrenia, *Nature*, 468:203–12.

Vayena, E. and Blasimme, A. 2017. Biomedical big data: new models of control over access, use and governance, *J Bioeth Inq.*, 14:501–13.

Wadhwa, P.D., Buss, C., Entringer, S. and Swanson, J.M. 2009. Developmental origins of health and disease: Brief history of the approach and current focus on epigenetic mechanisms, *Semin Reprod Med*, 27(5):358–68.

Warsh, J. 2014. PPI: Understanding the Difference Between Patient and Public Involvement, *Am J Bioeth*, 14:25–6.

Wartiovaara-Kautto, U., Hirvonen, E.A.M., Pitkanen, E., Heckman, C., Saarela, J., Kewttunen, K., Porkka, K. and Kilpivaara, O. 2018. Germline alterations in a consecutive series of acute myeloid leukemia, *Leukemia*. 32:228–85. https://doi.org/10.1038/s41375-018-0049-5

Weaver, I.C.G., Cervoni, N., Champagne, F.A., D'Alessio, A.C.V., Sharma, S., Seckl, J.R., Dymov, S., Szyf, M. and Meaney, M.J. 2004. Epigenetic programming by maternal behaviour, *Nature Neuroscience*, 7(8):847–54.

WHO (2014) www.who.int/end-childhood-obesity/commissioners/first-meeting-report/en/, Retrieved 14 July 2020.

Wicks, P., Vaughan, T.E., Massagli, M.P. and Heywood, J. 2011. Accelerated clinical discovery using self-reported patient data collected online and a patient-matching algorithm, *Nat Biotechnol*, 29:411–14.

Woolley, J.P., McGowan, M.L., Teare, H.J., Coathup, V., Fishman, J.R., Settersten, R.A., Sterckx, S., Kasye, J. and Juengts, E.T. 2016. Citizen science or scientific citizenship? Disentangling the uses of public engagement rhetoric in national research initiatives, *BMC Med Ethics*, 17:33.

Zhang, J., Walsh, M.F., Wu, G., Edmonson, M.N., Gruber, T.A., Easton, J., Hedges, D., Ma, X., Zhou, X., Yergeau, D.A., et al. 2015. Germline Mutations in Predisposition Genes in Pediatric Cancer, *N Engl J Med*, 373(24):2336–46.

Index

absolute risks 68, 135
accessibility 107, 152, 153
Adjekum, A. 149
adversity 164
agency 111, 142, 152
agent–object relationship 65
Alan Turing Institute 154
algorithms 16, 148–51, 153–5
Alzheimer's disease (AD) 19, 72, 85, 86, 136; prediction and prevention of 72–4
Angelina-Jolie-effect 68
appropriate-response interpretation 55–7
artificial intelligence (AI) 148
Austin, Jehannine 27, 28, 30
Ayo, N. 73

Barker, D.J. 163
Beauchamp, T.L. 52, 121
behavioural control 30, 31, 41–2, 44–5
Beier, K. 168
Belmont Report 63
biases 29, 135, 137, 138
big data 1, 18, 62, 149–51, 159, 168, 169; algorithms, transparency age 153–4; transparency and 151–3
bioethics 4, 5, 63, 168, 169
biomarker testing: for oncology 70–2
'Bio-Nespresso'-machine 20
'Black-Box' approach 153
bone marrow transplantation 51, 52, 55–60
Brasure, M. 73
breast cancer 13, 17, 62, 67–70, 85, 134, 135, 162; risk 8, 16, 134
BReast CAncer *(BRCA)* 13

cancer patients 17, 19, 72, 162
cardiovascular diseases 37, 136
care pathways 19
carrier testing 131
caveat emptor 119
Champagne, Frances 165
Chatterjee, N. 16
childhood obesity 165–8
children 14, 15, 19, 29, 71, 85–7, 90, 162, 167
Childress, J.F. 52, 121
citizens 6, 18, 19, 51, 53, 54, 58, 64, 132, 134, 139, 154, 160
clinical risk communication 51–61
clinical trials 20, 151–3
Coglianese, C. 154
cognitive control 41, 43–4
comity model 93
communicating risks 51, 60, 63
communication 2–4
conflicts 52, 69, 83, 152
consent 55, 58–60, 84, 94, 169
CRISPR 20

Daniels, Norman 160
data-intensive research 168, 170
decisional control 41–2, 44
decision-making 2–4
decision sciences 154–5
dementia 1, 72–4
deterrence 122
Digital Health Innovation Action Plan 150
direct to consumer personal genomic testing (DTC PGT) 8, 9, 101–25; companies 101, 105, 106, 108, 109, 111, 114, 118–20, 122, 124; cross-country comparison, main themes 118–19; data collection and analysis 104; data protection/privacy 118; Dutch participants 110–13; German

DTC PGT *continued*
participants 105–6; Israeli participants 107–10; lay perspectives, relevance of 103–4; major differences 121–5; major similarities 119–21; personal and clinical utility of 118; positive and negative social and moral issues 118; regulation and governance 119; United Kingdom participants 113–15
disease risk 37, 41, 46, 89, 90; estimates 118; information impacts 91; perceptions 89
DNA-based disease risk information 135
DNA-test 13–14, 16

electronic health records (EHR) 149
emotional regulation 6, 27, 31, 41–2, 45
emotional states 137, 138
empirical ethics 4–5
empowerment 6, 7, 27, 30–2, 35, 40–6, 133, 138–42; behavioural control 44–5; cognitive control 43–4; decisional control 44; emotional regulation 45; hope 45
engagement 30, 41, 168–70
environmental risk factors 35
epigenetic mechanisms 163–6
epigenetics 37, 159, 163–8
epistemic or causality dimension 66
ethical contexts: responsibility 63–6
ethical issues: of genetic and health risk 4–5
EuroGentest 121
European League Against Rheumatism (EULAR) 35
European Medicines Agency (EMA) 152, 153

familial hypercholesterolemia (FH) 13, 17–18
familial relationships 80, 81
familial responsibilities 80–2, 92–4
families 8, 14, 28–30, 65, 80–7, 91–4, 162, 167; genetic decision-making within 84–8; genetic risk information, handling 92–3; genetic testing and 82–92
family condition 27, 28, 30, 31
family members 7, 80–4, 86, 89–93, 132, 137, 138
family responsibilities 80–2

General Data Protection Regulation (GDPR) 152, 154
general public literacy 138–42
gene therapies 19, 20
genetic and biological risk profiling 36–42; empowerment 40–2; medicalization 36–40
genetic biomarkers 67
genetic counselling 6, 8, 14, 15, 27–31, 69, 130, 161; patients 27, 28, 30
Genetic Counselling Outcome Scale (GCOS-24) 6, 27, 31–2
genetic counsellors 13, 14, 27, 28, 69
genetic diseases 6, 15, 36, 37, 71, 82, 86, 87, 111, 115, 118
genetic information 6, 8, 40, 41, 79, 80, 82–4, 92–4, 118, 120, 161
genetic knowledge 8, 9, 14, 79, 135, 139, 140; cognitive aspects 137–8; of disease predisposition 135–8; and education 139–40; health beliefs, heuristics and biases 137–8; psychological aspects 135–7; technologies to foster 139–40
genetic responsibility 63–4
genetic risk factors 35
genetic risk information: public perspectives on management of 84–8
genetic risk perception 28, 30
genetic services 6, 13, 14, 93, 94
genetic testing/tests 29, 82, 84, 85, 87, 109, 123, 135, 136, 138, 142; family and 82–92; *pret-a-porter* 130–2
genetic unity 93
genomic health literacy 132–4
genomic medicine 5, 6, 13–15, 17, 19, 21, 131, 148; in 2025–30, 13–21; algorithms, policy tools 148–55; genes, comprehensive risk models 16; genomics based treatment 19–21; medical genetics, medical speciality 13; new ethical issues 14–15; non-genetic experts involvement 13–14; oncology 16–17
Germany 8, 87, 101, 102, 104, 105, 118, 121–3
Gigerenzer, G. 68
Gilbar, R. 94
Goold, S.D. 120–3
guilt 66

haematologists 7, 52–9, 61, 163
haematology 162
harmonization efforts 124

health care 1, 4, 6, 14, 17, 19, 84, 85, 121, 123, 167, 168, 170; professionals 14, 51, 52, 58, 59, 89, 123, 130, 159, 160; risk information in 1–2; services 79, 102, 137
health literacy 132–4
health numeracy 134–5
health-related DTC PGT 110, 113, 118, 119, 123
health responsibility 64
health risk communication 7, 62, 63, 67, 74; emergence in modern medicine 67–74
health risk information 5, 7, 39, 63, 67, 70
health risks 2, 4, 38, 62, 64, 69
Her2Neu 17
Hibbard, J.H. 41
high-risk genes 16
hope 42
Huntington disease 15, 62, 136

illness 36, 73, 74, 132, 136, 137
individual autonomy 69, 80, 93
individual patients 20, 69, 163, 170
individual responsibilities 57–8, 63, 64, 73, 160, 167
individual risks 16, 63
informational resource 42–4, 46
informational self-determination 122
inheritance 28, 29, 89, 138, 142
intimacy 81–2
Israel 8, 101, 102, 104, 107, 110, 119, 122, 123

Jonas, Hans 62, 63
Juengst, E.T. 40, 81

Kukla, Rebecca 58

learning 43, 140–2
Lehr, D. 154
Leonard, S.J. 94
leukaemia 51; patients 7, 51–3, 55, 57, 59, 61
Li Fraumeni Syndrome (LFS) 162
Lynch Syndrome 13, 17, 28–9

machine learning algorithms 148
mammography 68, 134, 135
markers of trustworthiness 121
material risks 52
Mayer, R.E. 141
McAllister, Marion 41
meaning: cognitive vs. evaluative 56
medicalization 7, 35–46
medical risks 51–61
Mesters, I. 133
Mill, Jonathan 164
'Million Genomes' project 18
Mind the Risk project 2, 46, 141–2, 159
Mirnezami, R. 154
models of inheritance 28, 29
monogenic cardiomyopathies 161
Montgomery v Lanarkshire 51
moral agents 65, 66
mortality risk ratios 166
Murgatroyd, Chris 164
mutations 14, 16, 17, 20, 68, 83, 135, 136, 142, 162, 163

National Health Service (NHS) 14, 15, 114, 118, 121
Nebula Genomics 101
Newson, A.J. 94
Next Generation Sequencing (NGS) 160–3
non-genetic experts 13, 14
norm–authority–consequences relationship, justification 65–6
Norwegian Strategy for Personalized Medicine in Healthcare 149
numeracy 131, 132, 134, 135
numeric probabilities 27, 28
nutrition 163–4

obesity 112, 137, 165–8
object relationship 65, 68
offspring 138, 165–7
oncology: biomarker testing for 70–2
O'Neill, Onara 59, 60

parental effects 165–7
Parikh, R.B. 151
particularism 4
patient activation 41
patient activation measure (PAM) 41
patient autonomy 52, 57–9
patient benefits 6, 31, 121, 161
patient communications 153, 154
patient–doctor relationship 18
patient empowerment 6, 27, 40, 41
patient engagement 41, 160, 168, 169
patient involvement 160, 170
patient regret 57–9
patient-reported outcome measures (PROMs) 30, 31

patient responsibility 57–9
Patients.Like.Me. 169
personalized medicine 35, 79–94
personal risk 7, 35, 138
pharmacological treatment 165
phenomenological interpretation 53, 57; risk understanding 53–5
philosophical-ethical usage 64
pleiotropy 131
policy making 3
polygenic risk scores (PRS) 6, 16–18
Potter, Van Rensselaer 63
precision medicine 35
predictive genetic tests 137, 138
pre-symptomatic diagnosis 15
preventive medicine 38, 39
professional responsibility 5, 7, 8, 62, 63, 67–9, 71–4; to avoid self-responsibilization 72–4
psychiatric disorders 164–5
'Public and Patient Engagement' 168–71
public education, genetics 141–2
public health communication 67, 68; women's breast cancer risk 67–70

Radio-Chemo-Therapy (RCT) 20, 71
rare diseases 5, 6, 14, 15, 168
Rawls, John 160
referential opacity 59–60
rheumatoid arthritis (RA) 7, 35, 37, 42–6, 88–90, 92; empowerment 43–5; managing risk of 88–92; medicalization 42–3; patient 37, 40, 46, 89–91; perspective of affected people 88–92; risk 46, 88, 90, 91
Rhodes, Rosamunde 81
Richards, M.P.R. 28
risk communication 1, 3, 51, 53–5, 57–60, 63, 91, 92, 154, 155
risk culture 28
risk empowerment 5, 6
risk figures 27, 28, 30
risk information 1, 2, 4, 6, 37, 42–6, 62–4, 71, 72, 74, 90–2; in health care 1–2; resource 45, 46
risk perception 27, 28, 30, 79, 81, 83, 85, 87, 89, 91, 93; accuracy 28; uncertainty and 79–94
risk profiling 35, 39, 40; genetic and biological 36–42
risks: difficulties in understanding 53, 55, 57, 59, 61; project 35, 46, 88, 132, 141; understanding 2–4, 7, 53, 55, 57, 59, 61; *see also individual entries*
Risling, T. 41
Rutten, Bart 164

Sabin, James 160
SARS-CoV-2 160
Schicktanz, S. 168
Schweda, M. 168
self-efficacy 132, 141, 142
Serious Games (SG) 141–2
side-effects 53–8, 60, 61
Single Nucleotide Polymorphisms (SNPs) 16
social environments 164
social groups 3, 65, 66, 73
social improvement 165
software as a medical device (SaMD) 148
Sparks, J.A. 40
Spengler, Dietmar 164
Strawson, P. 66

Theory of Justice 160
Triple negative Breast Cancer (TNBC) 163
trust 3, 65, 101–5, 113, 114, 119–24
trustworthiness 102–3, 121, 122
tumour 17, 71, 162, 163

Watson, D.S. 153
Wöhlke, S. 84, 85

Young, Iris Marion 81

For Product Safety Concerns and Information please contact our EU
representative GPSR@taylorandfrancis.com
Taylor & Francis Verlag GmbH, Kaufingerstraße 24, 80331 München, Germany

www.ingramcontent.com/pod-product-compliance
Lightning Source LLC
Chambersburg PA
CBHW060745150926
51501CB00003B/93
9780367356699